SPACE SCIENCE COMES OF AGE

Perspectives in the History of the Space Sciences

SPACE SCIENCE COMES OF AGE

Perspectives in the History of the Space Sciences

Edited by
PAUL A. HANLE
Acting Chairman of the Department of Space Science and Exploration, NASM

and

VON DEL CHAMBERLAIN
Astronomer, NASM

With contributions by
STEPHEN G. BRUSH
HERBERT FRIEDMAN
C. STEWART GILLMOR
LEO GOLDBERG
RICHARD P. HALLION
ROBERT JASTROW
PAMELA E. MACK
HOMER E. NEWELL
EUGENE SHOEMAKER
LYMAN SPITZER, JR.
JAMES A. VAN ALLEN

Published by the **National Air And Space Museum** Smithsonian Institution

Distributed by the **Smithsonian Institution Press** Washington City

1981

Library of Congress Cataloging in Publication Data
Main entry under title:

Space science comes of age.

Papers presented at a symposium held at and sponsored
by the National Air and Space Museum, Washington, D.C.,
Mar. 23-24, 1981.

Includes index.
1. Space sciences — History — Congresses. I. Hanle, Paul A.
II. Chamberlain, Von Del.
III. National Air and Space Museum.
QB500.S62   500.5'09   80.28966

ISBN 0-87474-508-X
ISBN 0-87474-507-1 (pbk.)

Endpapers: Montage of Saturn and four of its
moons—Dione, Tethys, Mimas, and
Enceladus—taken by Voyager I spacecraft in
November 1980. *(NASA photo.)*

This volume chronicling the emergence of
Space Science is dedicated to Dr. Thomas A. Mutch,
explorer of minds, mountains, moons, and
planets. Tim's vision and enthusiasm,
his skills as historian, researcher, teacher and
administrator, and his intellectual leadership
collectively form a pinnacle of performance
challenging all of us to attempt the climb.
Though he lost his earthly life among
the physical peaks of the beautiful Himalayas,
Tim's spirit remains with us,
leading us on to explore the universe about.

TABLE OF CONTENTS

The occasion for this book of essays recalling the origins of space science was a symposium with the same title held at the National Air and Space Museum on March 23 and 24, 1981. We wanted those days to bring together a distinguished panel of early contributors to the fields of space science, and to invite them to discuss among themselves, with historians, and our larger audience how they saw their field come into being. They were asked to evaluate its accomplishments in the quarter of a century or so of research in space and to put their accounts and evaluations down in these pages.

ACKNOWLEDGMENTS

The editors want to thank the contributors to this volume above all. Despite many demands placed on so distinguished a group, they submitted their works in good time and humor. The NASA History Office, especially Carrie E. Karegeannes, gave generous help with reprinting the essay by Homer Newell. We thank also the National Academy of Sciences for permission to reprint their transcription of James Van Allen's 1958 talk. The difficult job of editing the manuscript and placing all the works in a single format was accomplished with the invaluable aid of Peter Rohrbach. We thank Dorothy Fall, who designed the volume, and Maureen Jacoby, who managed its publication at the Press. Special thanks are due Lillian Kozloski and Carmen Smythe, who typed much of the manuscript. This publication and the associated symposium have enjoyed the enthusiastic support and guidance of Noel W. Hinners, Director of the National Air and Space Museum.

PREFACE

NOEL W. HINNERS

The symposium "Space Science Comes of Age" was conceived as a public debut for the National Air and Space Museum's concerted research effort to study the history of the "Space Age." This new direction augments already established programs in the collection, conservation, and exhibition of significant space-program artifacts and builds upon a base of prior studies in the history of rocketry. With this expansion of its charter, the Space Science and Exploration Department has a rare opportunity to fulfill the oldest of Smithsonian goals—"the increase and diffusion of knowledge among men"—with a program of high academic quality and focus of purpose normally hard to come by in the characteristically amorphous museum environment.

The very title of the symposium implies that "Space Science" and, indeed, the "Space Age" have not been considered valid topics for scholarly historic study. This is the view of some professional historians, who feel that the Space Age, if defined as starting with the Sputnik era of the late 1950s, is of too recent vintage to allow for proper historical perspective. To some extent this is true—it is indeed too early for us to comprehend the full significance of space exploration to future generations and no doubt their historians will be better judges of that if . . . . If they have adequate records of fact, and if they understand the political, social, economic, scientific, military and technological context of the times.

Think what better historical studies we would have now of the Wrights, Goddard, commercial aviation, military influence, etc. ad infinitum, if historians of the day had made more effort to acquire and preserve relevant documents and artifacts and if the "doers" better understood their obligation to save the evidence of how things happened. In the age of the telephone, too, it has become imperative to record their stories in forums such as this and with modern technological aids and oral-history interviews. Our immediate challenge and the impetus for getting started now on our new endeavor is to guarantee that in a century or two the history of space exploration will not be distorted for lack of evidence and historical sensibilities at its birth.

Noel W. Hinners is Director, The National Air and Space Museum

INTRODUCTION

PAUL A. HANLE

The dawning of the space age in 1957 did not mark the first time that humans undertook scientific study in space. Well before the launch of Sputnik I a small group of geophysicists and astronomers had begun experimenting above the atmosphere. With high-altitude balloons, German V-2 missiles, and new American sounding rockets, physical science took a great leap upward just after World War II. But it was the orbiting of the Russian and American satellites in 1957-8 especially that electrified the world of physics and brought home to many scientists the growing opportunities for research offered by these new laboratories in space. Along with military and political aims, science was central to the early space programs. It was a fact that we sometimes forgot in the clamor that surrounded the race to put a man on the moon.

The theme of this volume, if a single theme can emerge from a collection of contributions, is that space science—research in geophysics, astronomy, physics, atmospheric science, and life science which is devoted to understanding space and its contents—has come of age. By this we do not mean that we know all one can about space. On the contrary, scientists are only now approaching an understanding of how to undertake space research in a systematic way. By the standards of history, however, space science is beginning to develop a base of empirical knowledge, which can support an equally developing theoretical conception of space and its contents that *deserves* the toughest tests of future researchers. Not all of that conception will withstand the tests, as some contributors suggest. But it will be taken seriously, as it often was not a few decades ago.

There are other marks of maturity. The space sciences have their own source of funding, the NASA Office which is their great patron. They have political and scientific constituents, their own internal bureaucracy, and a variety of interdisciplinary programs of study, including several academic departments. These are indicators of an underlying community—a loose linking of activities and disciplines— that is necessary for people engaged in an intellectual pursuit to exchange information and so to reinforce the effects of each other's work. In spite of the tenuous nature of the community, it is an identifiable group; and with a little optimism one can see it as sustaining.

With the publication of this volume we would like to offer another measure of maturity. In the past few years professional historians have become interested in chronicling the development of the space fields. There is a pressing need to approach this history, and as soon as possible, because scientific research more than other space activities is subject to cannibalization of its artifacts, the instruments of detection and measurement. Because in science there is a professional value of suppressing the first person singular, and because the founders of the fields will not be here forever, we shall soon be loosing valuable written and oral evidence for the historians. To preserve this evidence and to present the history of the space sciences are tasks that the National Air and Space Museum has embraced with enthusiasm. We are pleased that as we do so the American Geophysical Union and the American Astronomical Society have offered their endorsement of this symposium. These organizations have agreed that it is important to show what has been achieved in the last 20 or 30 years and to present these historical insights to an audience larger than normally reads the professional literature of space science.

I think you will see in this volume how truly extraordinary and fascinating are the discoveries of space geophysics and space astronomy. What you will not see is a historical panorama drawn with broad brush-strokes such as we begin to see in the history of physics. Here is but a first step and will, I hope, form a part of the way upon which later analytical studies and larger syntheses will follow.

More than 20 years ago Robert Jastrow edited a collection of papers presented at a major symposium on space physics sponsored by the National Academy of Sciences, the National Aeronautics and Space Administration, and the American Physical Society. These works, several by contributors to this volume as well, were printed in 1960 under the title *The Exploration of Space*. A remarkable collection of scientific survey papers of a field at its inception, the volume still has relevance to present research. The collection was an introduction to the possibilities in space research; here we attempt instead to present a preliminary assessment of the results in following years. What is striking is that the basic outlines for what has been accomplished were drawn then; much of what we now know in detail was suggested then in its general form.

This fact is cited not to denigrate the enterprise, for it is just that detail—those magnificent photographs of the planets, the plethora of magnetospheric data, the X-ray and ultraviolet images of distant stars and of our own sun, for example—that will in time provide the basis for a greater scientific synthesis, sustenance for the inevitable grand shifts in scientific thinking that some observers call "scientific revolutions." In astronomy we have seen already our conception of the universe altered from the stable, relatively unchanging cosmos of 30 years ago to a violently active, evolving generator of the elements, the planets, the stars, and all else contained in its bounds. This change in our understanding of the universe has transpired in large part because of the discoveries of space astrophysics—discoveries which, like this volume, are little more than two decades old.

Our contributors, founders of the field of space science in Part I and professional historians in Part II, cover selected topics in early space astronomy and space geophysical sciences and a few aspects of the history of space physics. The essays are representative, not intended to cover all scientific areas but to give you overviews of selected fields of concentration from their own perspectives. The historical works are more diverse. What is not covered among them is clear at a glance. Historians have yet to address most of modern astronomy and planetary geophysics, mainly because of their newness. The historical essays here constitute practically all lines of present investigation in history of space sciences. Despite an occasional outstanding monograph, such as R. Cargill Hall's Project Ranger history, there seems so far to be very little historical interest in the questions of how space physics has evolved from its roots in cosmic-ray, astro- and geophysical research. It would be useful, for example, to investigate how James Van Allen's studies of the radiation belts of the earth were linked to earlier investigations of cosmic rays and the aurorae. As Dr. Van Allen has suggested, such links explain in part why he conducted experiments on the Explorer satellites, and fit well into a context of what earlier physicists wanted to know about the upper atmosphere. That context Stewart Gillmor has begun to examine, both here and elsewhere.

Stephen Brush broadly traces the development to 1960 of theories of the origin of the solar system. Brush acknowledges that thereafter the story becomes much more complicated but also more important for understanding where we now are in planetary science.

The technology of remote sensing was developed for photogeology of the planets, for space astronomy, and for looking down on earth's activities and resources. It would be interesting to see a technical study of the development of remote-sensing instrumentation in the future. Pamela Mack's essay, on the other hand, is a valuable overview of the politics and economics of one of the quasiscientific remote-sensing satellites. Some of her historical method might be applied as well to the basic-science research that forms the largest part of this book.

Richard Hallion's history of space-launch vehicles, with its insights into what drives technological development, treats one of the few topics in space technology to have received major historical treatment.

We are reprinting two essays from other sources. The original transcription of a talk that James Van Allen gave to the National Academy of Sciences in 1958, announcing the discovery of the trapped-radiation belts that bear his name, appears in essentially original form. It provides the historian and the layman with an example of how a recognized geophysicist presented this major piece of work to a group of his peers in 1958. The essay is offered as "raw material" for the historians to elaborate upon in coming years.

Homer Newell's essay gives us a comprehensive survey of the evolution of the space sciences in the 1960s, from the vantage of the man who administered those programs at NASA. The view is broad in scope, and emphasizes programs, thus providing a framework into which we may place several of the other contributions.

In these pages you will find a first look at the history of space science. We hope that it will stimulate historical discussion and study in a concerted way, to begin a serious treatment of history of space sciences in their scientific, social, political, and economic aspects. In astronomical and geophysical sciences it is physics that is the common denominator. We would like to see efforts comparable to those devoted to the history of modern physics directed toward preserving and chronicling its successor fields in space. In the history of modern science, as in physics, space affords exciting opportunities and insights.

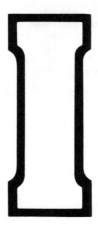

ASTRONOMY AND GEOPHYSICS IN SPACE

ULTRAVIOLET SPECTRA OF THE STARS

LYMAN SPITZER, JR.

etection and measurement of ultraviolet radiation from the stars has been a spectacularly successful field of space research. The atmosphere of the earth is very strongly absorbing for electromagnetic radiation with wavelengths shorter than 3000 A (angstroms). Hence such radiation cannot be detected from the earth's surface; telescopes above the atmosphere are needed.

It has been known for many years that measurement of stellar radiation at these ultraviolet wavelengths would provide data of very great scientific importance. Much of our knowledge of the stars and of interstellar clouds has been based on measurement of "absorption features" in stellar spectra; each feature is a narrow range of wavelengths in which electromagnetic radiation is selec-

tively absorbed by specific atoms. The absorption produced by each atom is centered at certain characteristic wavelengths, which are different for different atoms. Measurement of the central wavelengths for the absorption features in a stellar spectrum can indicate which atoms are present between the earth and a star. Remarkably enough, detailed measurements and theoretical analyses of these features can indicate which ones are produced in the atmosphere of the star and which are produced in the vastly more rarefied clouds between the stars. Such basic physical quantities as the temperature and density of the interstellar gas we can sometimes determine from the interstellar absorption features.

The importance of the ultraviolet spectrum is that it contains information not available from a measurement of visual or infrared light. For example, it turns out that most atoms and molecules in interstellar space do not absorb radiation at wavelengths between 3000 and 10,000 A. (An angstrom unit is 10^{-8} cm; the human eye is usually sensitive to radiation of wavelengths from

Lyman Spitzer, Jr. is Henry Norris Russell Professor of Astronomy in the Department of Astrophysical Sciences at Princeton University. He directed Princeton's development of the telescope for the third Orbiting Astronomical Observatory.

about 4000 to 8000 A.) In sharp contrast, most of these atoms and molecules will produce a variety of absorption features between 1000 and 3000 A. The abundant interstellar atoms,—hydrogen, carbon, nitrogen, oxygen, silicon and sulphur, and the abundant hydrogen molecules, in particular—can be detected in absorption at these ultraviolet wavelengths, but not at longer ones.

Another important characteristic of ultraviolet radiation is that a star of high surface temperature will emit most of its radiation at these short wavelengths. The radiation from the sun, at a surface temperature of about 6000° K, shows an intensity peak at about 5000 A. Stars with surface temperatures between 12,000° and 30,000° K, including about a third of the brightest stars in the sky, show the greatest radiated intensity (per unit wavelength) between 2500 and 1000 A. Measures in the ultraviolet are needed to find the total energy radiated from these stars, and to determine the temperature of the surface layers.

Many enthusiasts for rockets and space exploration have pointed out the value of space vehicles for ultraviolet research in astronomy. The early book by H. Oberth[16] in 1923 provided one of the earliest such discussions, with some professional astronomers also contributing over the years.[26, 21] It was not until after World War II, when rocket technology which had been developed during the war became available for space research, that the theoretical possibilities pointed out earlier began at last to seem almost practical. Some of the V-2 rockets developed by the Germans were brought into this country and became available for lifting scientific payloads well above the atmosphere for brief periods. A little later, smaller sounding rockets such as the Aerobee were developed in the U S to provide more flexible and, in time, more reliable vehicles for space research. After the launching of artificial earth-circling satellites during the International Geophysical Year,

1957-58, satellites gradually became available for a variety of scientific payloads.

I want to trace how the development of increasingly sophisticated space hardware made possible successive programs of ultraviolet stellar research, with steadily growing scientific capability. This discussion is not a general review of this field, but is mostly concentrated on those aspects of the U S program with which the author is most familiar.

Unpointed Sounding Rockets

When space astronomy became possible, the sun was naturally the first extraterrestrial target. The sun is brighter than the first-magnitude stars by a factor of about ten billion and can be measured with relatively simple equipment. In addition, to point an instrument at the sun is a relatively straightforward task; pointing at a star requires vastly more sophisticated engineering. As a result, while the first solar spectrum with sufficient resolution to show absorption features was obtained from space in 1946, research on night-time objects began only in 1957,[5] with substantial scientific results accumulating in the 1960s.

Until 1965 this early work was all confined to "unpointed" instruments on rockets. Figure 1 shows an Aerobee rocket, photographed during laboratory tests at the White Sands Missile Range, New Mexico. Once such a rocket was above 100 km, free of appreciable atmospheric drag, it would spin slowly about its cylindrical axis, with some wobble ("nutation") also. These rotational motions were typically unplanned and uncontrolled. The ultraviolet detectors flown on these rockets were normally either photomultipliers or Geiger counters used in a photon counting mode; each detector was sensitive over a wavelength band typically about 100 A wide. Each detector would count photons from a substantial area of the sky, sweeping around as the rocket rotated and tumbled. The output of the detectors was transmitted to the ground for recording

3

Figure 1. Aerobee sounding rocket in test stand at the White Sands Missile Range. The scientific instrument, in this case an ultraviolet spectrophotometer, is in the cylindrical section next to the nose cone.

and subsequent analysis.

A central problem with this type of instrumentation was to determine the precise orientation of the rocket at each moment during its flight. In principle this information can be obtained by the use of several photometers sensitive to visual light. As these sweep by stars of known position and visual brightness, the photometric signals can be used to determine the orientation of the rocket as a function of time. In theory there will be one and only one solution which is consistent both with the signals from the visual tubes and with the dynamical properties of the passive rocket structure. In practice this solution is difficult to find, and other clues were generally used to limit the range of orientation possibilities. Clues sometimes used were on-board magnetometer measurements or a determination of the horizon by means of photon detectors sensitive to air glow emission. With these techniques, identification of ultraviolet signals with known hot stars (spectral types O and B) was generally possible.

London showed that the ultraviolet spectra of hot stars were in reasonable agreement with theoretical models. In making this comparison two effects had to be taken into account. First, the presence of strong absorption features, which crowd rather closely together in the ultraviolet, was considered; because of this "blanketing" effect, less radiation escapes from a star in the ultraviolet than would otherwise be the case. Second, the presence of dust clouds between the earth and a star can absorb ultraviolet radiation more than visible light, altering the shape of the observed spectrum; prior knowledge of the amount of dust in the line of sight to each star (determined from the shape of the spectrum in visible light) helped in correcting for this effect.

Evaluation of this dust absorption in the ultraviolet provided a second major result from these unpointed rocket spectrometers. With decreasing wavelength the absorption was found to increase more rapidly than expected[3] and, moreover, showed a peak[25] at about 2200 A, such as might be produced by graphite particles. The subsequent detailed observations of these effects by the OAO-2 satellite (see below) made it possible to determine the possible composition and size distribution of interstellar dust particles.

Pointed Sounding Rockets

It was generally evident that the full potentiality of ultraviolet research on stars urgently required a pointing system for the instruments launched into space. To look at faint stars or to measure stellar spectra with moderate resolution (1 A or better) required longer exposures than could be obtained with an unpointed rocket, whose rotation tended to blur the slitless spectra in a very short time. While the Orbiting Astronomical Observatories Program, started in 1960, was devoted to the development of large, accurately pointed satellites (see below), the development of a pointing system for sounding rockets was also desirable; these smaller rockets provided a flexibility for new types

While the spectral resolution of such measurements was much too low to show any absorption features, wide-band measurements of stellar spectra were of great scientific interest. When the instrument had been correctly calibrated (and when the calibration did not change during the launch) such data gave the true shape of the stellar continuous spectrum, providing an important check on theories of stellar atmospheres and how much atmospheres radiate energy.

A survey[27] of about a dozen papers in this field which had been published by groups at the US Naval Observatory, the Goddard Space Flight Center, and the University of

of observations that could not be obtained with large satellites, with their necessarily long lead times. Accordingly, the Goddard Space Flight Center undertook to develop hardware for this purpose.

The resultant Attitude Control System (ACS), which was flown on Aerobee rockets as early as 1964, utilized jets of helium to orient the entire rocket, with small control gyros to control the jets. This system used the "bang-bang" technique, with the rocket drifting back and forth in a limit cycle, its motion reversed at each end with a small pulse of gas. In early operation the ACS oriented an Aerobee rocket with an accuracy of 1° to 2°, and held it steady within a limit cycle of about 0.2°.

This stability, while a great improvement over unpointed rockets, was not adequate to permit the spectral resolution of 1 A needed for measuring absorption features in stellar spectra. To improve this stability a large passive gyro was used, mounted in gimbals with the spectrograph so as to prevent rotation of the instrument in the direction of the spectroscopic dispersion. This system, developed and used by a group at Princeton University, had been suggested informally by J.E. Kupperian at Goddard. In the first successful flight, in 1965, the passive gyro held the spectrograph steady to about ±16 arcseconds in the dispersion direction, providing a wavelength resolution of about 1 A. The spectra of two stars were recorded on film, which, on development, showed[15] many absorption features between 1300 and 1600 A observed for the first time in a star other than the sun.

It is of interest to note some of the practical difficulties that beset this program, since these are reasonably typical of early space research efforts. The successful flight had been preceded by two unsuccessful ones. In the first flight, a change in winds blew the rocket off course, and the motor was turned off early; the residual fuel became mixed with the helium used for the ACS, clogging the jets, and no scientific data were obtained.

In the second flight the rocket trajectory was satisfactory, but an electronic failure in the ACS prevented any stabilization whatever. In both these flights, the instrument was ejected from the rocket and successfully brought down by parachute, with only minor repairs needed for a subsequent flight. In the third flight, all mechanical and electrical systems seemed to operate properly. However, the parachute failed, and the instrumentation was smashed beyond repair. The film magazines had partially broken open, and lay in the desert sun for about an hour before they were found. It was a great surprise, and an unexpectedly pleasant one, when careful underdevelopment of the films two weeks later showed useful spectra of two bright stars in the constellation Scorpius. Other rocket research groups had similar experiences, alternating between sorry failures and reluctant successes.

Similar equipment was used for some dozen flights during the next few years; about half of these flights gave useful scientific results—about the average success ratio at that time. One of the most interesting scientific results obtained was the discovery of rapid mass loss from hot supergiant stars. The profiles of some of the strongest absorption features in these stars indicated that hot gas must be flowing away from these stars at velocities of some 1000 to 2000 kilometers per second. This velocity is indicated by a large shift of these absorption features toward shorter wavelengths, a shift interpreted as due to the familiar Doppler effect of a moving light source. This effect has been fully confirmed with sounding-rocket studies by other groups and, subsequently by studies with the OAO-3 satellite, *Copernicus*. A detailed survey[20] showed that most of the brighter hot stars are losing mass, an effect which greatly modifies our understanding of a star's evolutionary history.

Spectrometers on pointed rockets began observational studies of emission features in cool supergiant stars.[14] The strong hydrogen emission which was observed at 1216 A and which indicates chromospheric activity in such objects was extensively observed in later years with satellite instruments.

Pointed sounding rockets were also used for research on the interstellar gas, through measurement of the absorption which this gas produces in ultraviolet light from a star. In particular, a flight[6] of a Naval Research Laboratory spectrograph in 1970, using a special windowless detector developed for this purpose, detected for the first time the strong ultraviolet absorption features of interstellar molecular hydrogen along the line of sight to one bright star. Modifications in the rocket pointing system reduced the limit cycle substantially and permitted spectroscopy at moderate resolution (about 2 A) without the use of such additional features as the passive gyro developed for the Princeton flights. Since hydrogen is overwhelmingly the most abundant gas in the Universe (except in such atypical spots as the earth) the discovery that much of this element is in molecular form in interstellar space has had profound effects on our understanding of physical processes between the stars. The important field of research opened by this discovery was extensively exploited a few years later by the *Copernicus* satellite, which, starting in 1972, supplemented sounding rockets for research on ultraviolet absorption features.

Astronomical Satellites

The launch of the USSR Sputnik, on October 4, 1957, greatly accelerated the U S program of space research. NASA was organized in late 1958, with a broad mission for an energetic space program. In March 1959, NASA set up a Space Sciences Working Group on Orbiting Astronomical Observatories (OAO), whose membership included a number of astronomers interested in astronomical research with artificial earth satellites. The OAO program which grew out of discussions with this and other groups included the development (by Goddard) of a standard spacecraft, of which

several would be built and launched, each carrying a different astronomical instrument. The spacecraft included guidance on bright stars (supplemented later with a more flexible and reliable inertial guidance system), solar paddles, batteries, suitable memories for commands and for data—in general, the by now quite familiar panoply of engineering items needed for each space vehicle. Requests for proposals for the spacecraft and for the science instruments were issued in 1960 and 61. Figure 2 depicts a "typical" OAO spacecraft, while Figure 3 indicates a cutaway version of the spacecraft, showing the location of astronomical instruments in the central tube, 40 inches in diameter and 10 feet long.

In view of the pioneering character of this program, which marked an enormous increase in the sophistication and flexibility of space hardware, it is not surprising that difficulties arose with a number of technical problems, especially those associated with use of high voltages in a vacuum environment and precise guidance of the spacecraft. These problems were all solved in due course, but with some delay in schedule and with considerable increase in cost over the initial estimates. In the course of the program, four OAO spacecraft were launched. One of these failed to reach orbit, because of a malfunction of the hardware for separating different rocket stages. Of the other three, OAO-1 died after a few days in orbit, because of a battery failure, perhaps resulting in part from high voltage arcs and the effects which resultant electrical transients produced on the control circuitry. The other two satellites, OAO-2 and OAO-3, which were launched on December 7, 1968, and on August 21, 1972, were fully successful, and yielded fascinating and important scientific results for four and eight years, respectively.

OAO-2 contained two sets of instruments, one looking out each end of the octagonal spacecraft. The University of Wisconsin provided a group of photometers and low-resolution scanning spectrometers, while the

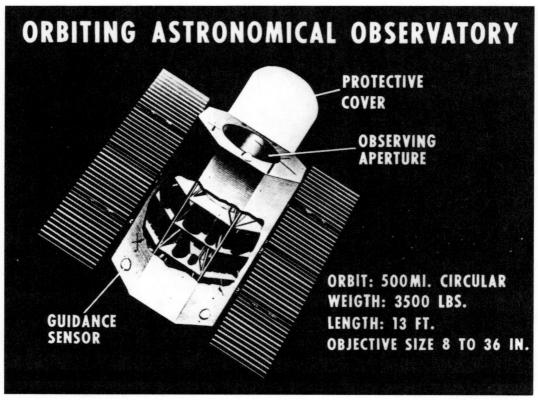

ORBITING ASTRONOMICAL OBSERVATORY

PROTECTIVE COVER

OBSERVING APERTURE

GUIDANCE SENSOR

ORBIT: 500 MI. CIRCULAR
WEIGTH: 3500 LBS.
LENGTH: 13 FT.
OBJECTIVE SIZE 8 TO 36 IN.

Figure 2. Sketch of an Orbiting Astronomical Observatory, showing the solar paddles, the guidance sensors (each a small star-tracking telescope), and the interior bays for engineering equipment. The protective cover, when open as shown, serves as a shade to keep sunlight out of the observing aperture.

Smithsonian Astrophysical Observatory provided a group of four 12-inch telescopes, each with its own wavelength band and each with a small television camera (a Westinghouse Uvicon) for recording an image which was subsequently transmitted down to the ground. As part of the overall project plan, the precise guidance required for the last OAO was not provided on the earlier models; hence OAO-2 was guided entirely by its star trackers, which gave a pointing accuracy of about one arcminute, and a pointing stability of about one arcsecond. The scientific programs of OAO-2 were in large part directed towards wide-band photometry; in addition, the 12 A resolution of the scanning spectrometer made possible measurements of the widest absorption features, particularly the one at 1216 A (designated as Lyman-alpha) produced by interstellar hydrogen atoms. The extensive scientific results[1,7] from OAO-2 were summarized in a special symposium.[8]

Two European satellites at about this time carried equipment for ultraviolet measurements of stellar spectra. Each was launched into polar orbit from the Western Test Range in California. The TD-1 satellite, launched March 12, 1972, contained two ultraviolet spectrophotometers. One was a joint Belgian-UK instrument[4], somewhat similar in purpose to OAO-2; this instrument contained a photometer measuring light at 2740 A, in a band some 300 A wide, and a three-channel grating spectrometer operating from 1350 to 2550 A, with measuring band widths of about 40 A. The other

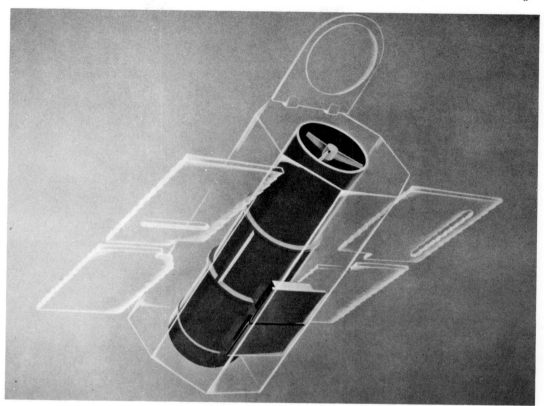

Figure 3. Cutaway diagram of an Orbiting Astronomical Observatory, showing a telescope placed in the observing aperture. The structure shown at the upper end of the telescope is the secondary mirror and its supporting structure.

spectrophotometer on the TD-1 satellite was a moderate-resolution Dutch instrument[12] which scanned three 90 A spectral ranges between 2000 and 2900 A; the measuring band width was 1.8 A. The Astronomical Netherlands Satellite, launched August 30, 1974,[28] had five measuring bands between about 1800 and 3300 A, with widths between 150 and 200 A. Each of these three instruments functioned for about two years, and carried out research on a variety of astronomical problems. Several of the U S manned satellites in the late 1960s and early 1970s also carried important ultraviolet instruments.[11,9]

The first astronomical satellite with a capability for high spectral resolution was OAO-3, which after launch was named *Copernicus,* in honor of the 500th anniversary of the great Polish astronomer. For this telescope-spectrometer, provided by the Princeton University Observatory, precise guidance was required. Starlight reflected off the jaws of the spectrometer entrance slit was used to provide an error signal; the fine-guidance system, including this signal, several rotating wheels, and appropriate electronics, was able to rotate the spacecraft to bring the star image on to the spectrometer entrance slit and to hold it there with an accuracy of 0.02 arcseconds for periods of several minutes. The success of this guidance system represents a very significant engineering achievement, which helped pave the way for the Space Telescope and the even more precise guidance that it requires.

The *Copernicus* instrument[18] comprised a Cassegrain telescope, which formed a stellar image on the entrance slit of an ultraviolet spectrometer. From the standpoint of research performance the two really essential properties of this spectrometer were: (a) its high resolution, about 0.05 A at 1000 A; (b) its high sensitivity for wavelengths between 950 and 1130 A, achieved with windowless photomultiplier tubes and with LiF coatings on the three reflecting surfaces (32-inch primary mirror, secondary mirror, concave grating). The three particularly significant *Copernicus* results which are described below were all based on observations at these relatively short ultraviolet wavelengths. Since the instrument was designed to count the photons detected at each wavelength during an "integration" time of 14 seconds, with the phototube then moving 0.025 A for the next integration period, appreciable time was required to scan even a single absorption feature. Nevertheless, during its eight years of operation, *Copernicus* obtained a large amount of data. Figure 4 shows a scan over 8 A of the spectrum of the bright reddened star ζ Ophiuchi. The error bars shown for two of the absorption features (at 1048.2 and 1053.3 A) represent the statistical fluctuations of the photon counts; the overall photometric errors were normally close to these theoretical values.

Copernicus observations have had a pro-

found effect on our knowledge of the rarefied material between the stars.[24] Three particular results may be mentioned. First there is the discovery that much of the interstellar hydrogen is in molecular (H_2) rather than in atomic (H) form. Research on this topic had been one of the major goals of the Princeton instrument, which had been optimized for wavelengths less than 1130 A, since all the strong H_2 absorption features have shorter wavelengths. While the first discovery[6] of interstellar H_2 had been made with a sounding rocket two years before the *Copernicus* launch, the high resolution and the 24-hour-a-day observing capability of the *Copernicus* spectrometer were required to establish[24] the widespread distribution of H_2. In addition, the satellite data, which gave precise measures of weak H_2 absorption features as well as of strong ones (see Figure 4), made it possible to use H_2 molecules as a measuring probe of the interstellar gas,

Figure 4. Spectrum scan of the bright star ζ Ophiuchi obtained with the *Copernicus* satellite. The plot shows the number of photons counted during a 14-second interval at each of a series of wavelengths, separated from each other by 0.025 A. The wavelengths of absorption features produced by neutral argon atoms (ArI) and by H_2 and HD molecules in various excited states (of vibration and rotation) are shown by vertical lines.

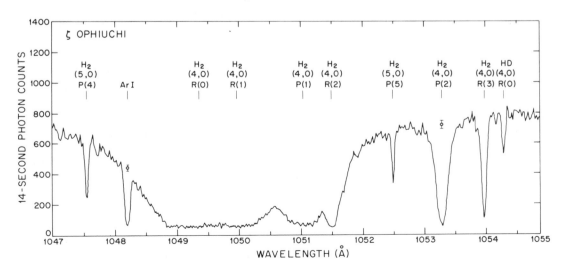

yielding values for the density and temperature within interstellar clouds. We now know that there is at least as much H_2 as there is H in clouds throughout the Galaxy, and that the temperature of a typical cloud studied by *Copernicus* is about 80° K, or, on the familiar Fahrenheit scale, about −315 degrees. The presence of so much molecular gas has far-reaching effects on the physics and chemistry of the interstellar material, greatly affecting, for example, the condensation of interstellar clouds to form new stars.

A second result referred to here is the presence[24] of a very hot gas in some regions of interstellar space. I had hypothesized the existence of such a hot component of the interstellar gas some 20 years earlier,[22] and to search for such a hot gas had been one of the Princeton objectives. It turns out that an oxygen atom that has lost five of its electrons absorbs strongly in the ultraviolet at about 1031.9 and 1037.6 A; to produce such highly ionized atoms by collisions in a gas requires a temperature of about a million degrees K. This pair of absorption features was in fact found to be present in most stars, and a detailed analysis[10] shows that the hot gas is widely distributed, not concentrated around a few hot stars. Soft X-rays are also observed from this hot gas, an effect which now confirms a range of temperatures centered at roughly a million degrees for this component of the interstellar material. To understand the origin, motions, and evolution of this hot gas and its influence on the Galaxy as a whole provides a major challenge for future astrophysicists.

A third result is the relative abundance of the two hydrogen isotopes. An ordinary hydrogen atom consists of a proton together with one orbital electron. An atom of deuterium consists of a neutron and a proton bound together in the nucleus, with the same electric charge as the proton but twice the mass, again accompanied by one orbital electron. The absorption features produced by these two atoms have wavelengths which differ by about 0.3 A; at wavelengths between 938 and 1025 A four different pairs of absorption features produced by these two isotopes have been measured separately by *Copernicus*.[17] The data indicate[13] a number ratio of deuterium atoms to ordinary hydrogen atoms equal to about 1.5×10^{-5}. The importance of this measurement is that the present value of this ratio is believed to depend primarily on conditions during the so-called "Big Bang," when the Universe was created in a gigantic explosion. If this theory is accepted, the measured deuterium-hydrogen ratio tells us, after considerable analysis, that the present smoothed density of the universe is 5×10^{-31} gm/cm^3, one tenth of the value required to close the universe and ultimately to reverse the present expansion. The argument is by no means watertight, but certainly provides strong support for the view that the universe will go on expanding indefinitely.

Many other observational programs were also carried out by *Copernicus*, some in the field of interstellar matter,[24] others[19] in a variety of stellar and planetary research areas. About half of the telescope time was used by some 200 Guest Investigators from 14 nations. From 1974 through 1979 about 40 papers based on *Copernicus* data were published each year.

The Present and the Future

I cannot close this review of ultraviolet astronomy without at least a brief glimpse at the present developments and at the future which can be expected to flow from the past.

At the present time and for the next few years research in ultraviolet astronomy is almost entirely concentrated in the International Ultraviolet Explorer (IUE). NASA planned to discontinue operations with *Copernicus* on February 15, 1981; since the sensitivity at short ultraviolet wavelengths has decreased steadily over the years, and by the end of 1979 had fallen to about 5 percent of its initial value, we could not strongly object to this course of action. The IUE instrument,[2] developed jointly by NASA and the

European Space Agency, uses a television tube to record an entire stellar spectrum in a single exposure, enormously increasing the rate of data acquisition, and extending research to much fainter stars and to some of the brighter extragalactic objects. While IUE does not have the high photometric accuracy or the short wavelength (below 1130 A) capability of *Copernicus,* its own special characteristics have opened new fields of ultraviolet astronomy. These fields are being energetically pursued by large numbers of astronomers in this country and in Europe. We may hope that this powerful equipment, launched on January 26, 1978, will operate for several more years.

Several years in the future is Space Telescope, a powerful man-maintained observing facility in orbit. With its 2.4-meter primary mirror, its two ultraviolet spectrographs (one with high spectral resolution, one with low) its two cameras, each equipped with ultraviolet filters, its high-speed photometer and its capability for precise astrometry, this magnificent equipment will be a uniquely powerful tool for astronomical research at ultraviolet wavelengths longer than 1130 A. The history of this project, including its gradual acceptance first by astronomers, then by NASA and finally by Congress, forms a story[23] of its own, which is outside the scope of this essay.

The one conspicuous gap in the capability of the Space Telescope for ultraviolet astronomy (defining this as the range from 900 to 3000 A) is the region at shorter wavelengths, from 900 to about 1130 A. This region contains important absorption features produced by H, H_2, singly and doubly ionized C and N, and O atoms ionized five times. Efficient detection of radiation at these wavelengths requires coating the optical surfaces with LiF, which is subject to degradation in the presence of water vapor or of various other contaminants. Photon detector tubes without windows are required for high efficiency, and these are also subject to contamination, an effect responsible in part for the gradual decay of the *Copernicus* sensitivity. Because of these practical problems it was decided not to include this short-wavelength capability in the Space Telescope, which will have quite enough technical problems in any case. However, a special satellite telescope designed for these short wavelengths, recording an entire stellar spectrum from 900 to 1130 A in a single exposure, would be a major facility for future astronomical research. Such an instrument would permit extending to much greater distances, including nearby galaxies, some of the key interstellar research programs which *Copernicus* has shown to be so revealing.

AKNOWLEDGMENT

It is a pleasure to acknowledge helpful suggestions on these historical topics from E.B. Jenkins, D.C. Morton, and C.C. Wu.

REFERENCES

1. R.C. Bless and A.D. Code, *Ann. Rev. Astron. and Astroph.*, *10*, 197, 1972.

2. A. Boggess et al., *Nature*, *275*, 372 and 377, 1978.

3. A. Boggess and J. Borgman, *Ap. J.*, *140*, 1636, 1964.

4. A. Boksenberg et al., *M. N. R. A. S.*, *163*, 291, 1973.

5. E.T. Byram, T.A. Chubb, H. Friedman and J.E. Kupperian, *A.J.*, *62*, 9, 1957.

6. G.R. Carruthers, *Ap. J. (Lett.)*, *161*, L81, 1970.

7. A.D. Code and B.D. Savage, *Science*, *77*, 213, 1972.

8. A.D. Code, ed. *The Scientific Results from OAO-2*, NASA SP-310, 1972.

9. K.G. Henize, J.D. Wray, S.B. Parsons, G.F. Benedict, F.C. Bruhweiler and P.M. Rybski, *Ap. J. (Lett.)*, *199*, L119, 1975.

10. E.B. Jenkins, *Ap. J.*, *220*, 107, 1978.

11. Y. Kondo, K.G. Henize and C.L. Kotila, *Ap. J.*, *159*, 927, 1970.

12. H.J. Lamers, K.A. van der Hucht, M.A.J. Snijders and N. Sakhibullin, *Astron. and Astrophys.*, *25*, 105, 1973.

13. C. Laurent, A. Vidal-Madjar and D.G. York, *Ap. J.*, *229*, 923, 1979.

14. H.W. Moos and G.J. Rothman, *Ap. J. (Lett.)*, *174*, L73, 1972.

15. D.C. Morton and L. Spitzer, *Ap. J.*, *144*, 1, 1966.

16. H. Oberth, *Die Rakete zu den Planetenräumen* (R. Oldenbourg Verlag: Munich, 1923).

17. J.B. Rogerson and D.G. York, *Ap. J. (Lett.)*, *186*, L95, 1973.

18. J.B. Rogerson, L. Spitzer, J.F. Drake, K. Dressler, E.B. Jenkins, D.C. Morton and D.G. York, *Ap. J. (Lett.)*, *181*, L97, 1973.

19. T.P. Snow, *Earth and Extraterrestrial Sciences*, *3*, 1, 1976.

20. T.P. Snow and D.C. Morton, *Ap. J. Supp.*, *32*, 429, 1976.

21. L. Spitzer, *Astronomical Advantages of an Extraterrestrial Observatory*, Project RAND Report, Douglas Aircraft Co., Sept. 1, 1946.

22. L. Spitzer, *Ap. J.*, *124*, 20, 1956.

23. L. Spitzer, *Quart. J. Roy. Astr. Soc.*, *20*, 29, 1979.

24. L. Spitzer and E.B. Jenkins, *Ann. Rev. Astron. and Astroph.*, *13*, 133, 1975.

25. T.P. Stecher, *Ap. J. (Lett.)*, *175*, L125, 1969.

26. J.Q. Stewart, Lecture to Brooklyn Academy of Science, April 11, 1929.

27. A.B. Underhill and D.C. Morton, *Science*, *158*, 1273, 1967.

28. R.J. van Duinen, J.W.G. Aalders, P.R. Wesselius, K.J. Wildeman, C.C. Wu, W. Luinge and D. Snel, *Astron. and Astrophys.*, *39*, 159, 1975.

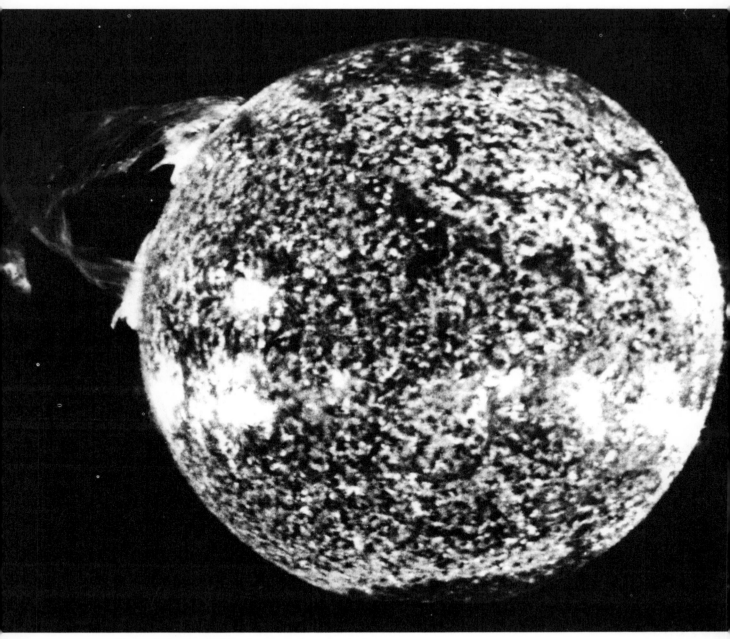

Figure 1. A spectacular solar flare photographed
December 19, 1973, by the Skylab 4 crew.
This is an extreme ultraviolet spectroheliogram
in the light of ionized helium.

SOLAR PHYSICS

LEO GOLDBERG

cientific interest in the observation of the solar spectrum from outside the earth's atmosphere began to heighten in the late 1920s, after advances in quantum mechanics and statistical mechanics opened the door to the quantitative analysis of the solar spectrum. It soon became apparent that the interpretation of eclipse spectra—made when the moon blots out the bright disk of the sun and exposes to view the relatively feeble radiation of the sun's outer atmosphere— would continue to be a semi-speculative exercise without knowledge of the intensities of the resonance lines emitted by the light and abundant elements, such as hydrogen, helium, carbon, nitrogen, oxygen, and others. To explain the intensities of lines in the visible spectrum, a theorist had only to invoke an appropriate, arbitrary model of the invisible far ultraciolet radiation field and no one could prove him wrong. Atmo-

Leo Goldberg is Astronomer, former Distinguished Research Scientist and Director of Kitt Peak National Observatory in Tucson, Arizona.

spheric ozone, oxygen and nitrogen were known to be the principal absorbers of solar ultraviolet light; ozone alone was recognized as the source of absorption between 3000 and 2200 A (angstroms). Since it was also known that ozone forms by the action of sunlight on oxygen, the Norwegian astrophysicist S. Rosseland[13] tried in 1929 to observe the sun in the north polar regions during the winter, when sunlight is weakest. Unfortunately, ozone in the polar regions was actually found to be more abundant in the winter[3] because sunlight both creates and destroys ozone. Instead of going north, one has to ascend to overcome the ozone absorption. Thus in Germany, the Regeners,[11] father and son, carried a solar spectrograph in a balloon to a height of 31 km and extended the solar spectrum about 100 A beyond the limit observed on the ground.

One of my most memorable experiences as a graduate student at the Harvard College Observatory was listening to a colloquium by the famous Indian astrophysicist M.N. Saha,[14] in 1937, when he suggested that a spectrograph be flown in a balloon to a height of 50 km and predicted that at this altitude the atmosphere would be transparent to UV radiation as far as 2200 A and might even transmit radiation through the band 1100-1250 A, which includes the revealing Lyman-alpha line of hydrogen. Even 40 km was a bit beyond the reach of balloons at that time, and moreover Saha's estimate of the height of the ozone layer was on the optimistic side.

In those days, most astronomers took the conservative view that the sun was an object in pure radiative equilibrium at a surface temperature of about 6000°K. The most prominent dissenter from this orthodoxy was Donald H. Menzel[9] of Harvard University, who was then engaged in quantitative studies demonstrating that the intensities of the Balmer series of hydrogen in eclipse spectra could not be explained unless the temperature were 10,000° K and that the helium lines required even higher temperatures up to 20,000° K. But the outstanding mystery of the solar spectrum lay in a group of about 20 unidentified emission lines, which were radiated by the corona and observed during eclipses.

In 1942, solar physics was shaken by B. Edlen's[4] discovery that the coronal lines are radiated by very highly ionized atoms, proving that the temperature of the corona is on the order of a million degrees. At once it became clear that the hydrogen and helium lines studied by Menzel were being radiated in the transition zone between the relatively cool "surface" layers and the very hot corona. The temperature rise from photosphere to corona is very steep and occurs in a very thin transition layer. Study of the structure of this transition region is essential to the understanding of the detailed mechanisms that heat the corona. Even now this problem is not completely solved, although we believe that the mechanism must be the conversion of mechanical and magnetic energy into thermal energy. Atoms in the corona are observed to radiate in very high stages of ionization in which 6 to 10 or even more electrons have been stripped away, whereas in the photosphere the atoms are neutral or singly-ionized at most. The transition region radiates emission lines from intermediate stages of ionization and since the degree of ionization increases with height, the observation of those intermediate stages, all of which radiate in the far ultraviolet, provides a probe for the elucidation of the structure of the atmosphere.

Once World War II was over, Edlen's revolutionary discovery provided the major scientific justification for putting solar telescopes above the earth's atmosphere. Even in 1942 astronomers had no expectation of observing the solar ultraviolet in the foreseeable future. The idea of lifting a spectrograph to an altitude of 100 miles or more, pointing it accurately at the sun and returning the data to earth seemed like science fiction. But under the impetus of war, the means became available when in 1944 the V-2 rockets were built to be launched from the European continent to strike London and other allied cities. It was immediately obvious to physicists and astronomers that, when pointed straight up and bearing instruments instead of bombs, these rockets could easily penetrate the ozone layer. Components from several hundred V-2's were captured by the US in 1945, and as soon as the war ended plans emerged to assemble a few score of complete rockets and use them for exploration of the upper atmosphere and for solar observation.

The scientific possibilities presented by the V-2 rockets certainly ignited my own personal interest in space research and my association with the McMath-Hulbert Solar Observatory turned my attention to the sun. During the spring and summer of 1946, Lyman Spitzer and I had several discussions concerning the possibility of our collaborating in establishing a laboratory for high altitude spectroscopy at Yale University with Office of Naval Research sponsorship. As it turned out, other responsibilities claimed our attention but from that time on I knew my future commitment to space astronomy was assured. I especially remember one day in July 1946 when Lyman and I sat on a park bench in Washington and jotted down ideas for solar space experiments which he incorporated into a more general memorandum on "Astronomical Advantages of an Extra-Terrestrial Observatory."[8] Among the observations we proposed were spectroscopy of the whole sun with a small ultraviolet spectroscope, and as a sequel experiment the ultraviolet spectroscopy of details on the solar surface such as sunspots, active regions, and prominences, using a 10-inch reflecting telescope and accessory equipment. Spitzer's memorandum was prepared for the late Dr. David T. Griggs as an appendix

to a study by the RAND division of the Douglas Aircraft Company of the feasibility of building, launching, and operating a small satellite observatory. The study concluded that although technically feasible the project should not be undertaken for two reasons: because of its high cost, which was estimated at two billion dollars; and because the lifetime of the satellite would be extremely uncertain in the absence of accurate knowledge of upper air densities.

Shortly after the Spitzer memorandum was published in September 1946, the first

Figure 2. NASA Goddard Space Flight technicians check out the first of the orbiting Solar Observatory Spacecraft prior to mating on the three-stage Delta rocket, February 1962.

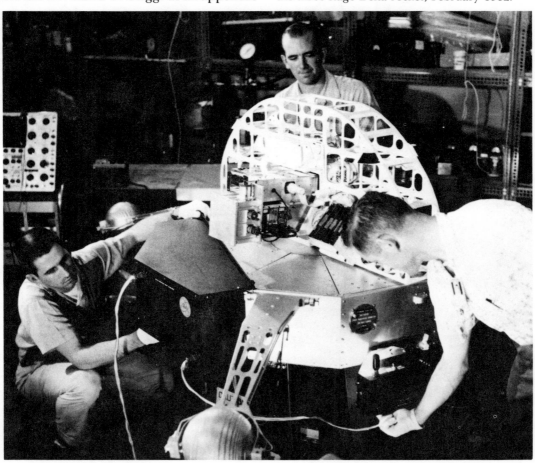

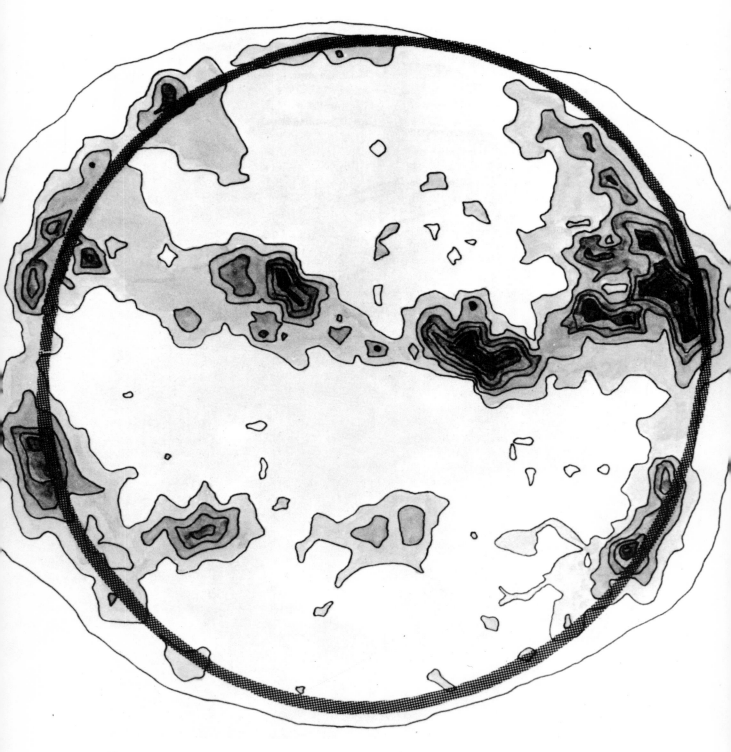

photographs of the ultraviolet spectrum were taken from a V-2 rocket instrumented by the Naval Research Laboratory.[2] During the ensuing period of 11 years that elapsed before the first earth-orbiting satellite was launched, a brilliant series of rocket experiments explored the solar spectrum from the atmospheric ultraviolet cut-off wavelength down to and including the region of soft X-ray emission.[5] The principal institutions engaged in this effort in addition to NRL were the Air Force Cambridge Research Laboratory, the University of Colorado and the Applied Physics Laboratory of Johns Hopkins University. Elsewhere in this book you will read the account of Herb Friedman, who was one of the true scientific pioneers of this era, and I shall therefore not venture into territory that bears his mark so indelibly. I will turn instead to the beginnings of solar research with satellites.

The story of the launching of the first artificial satellite is a familiar one. As a commitment to the International Geophysical Year, 1957-1958, the USA and the USSR undertook to launch small artificial satellites, primarily for the purpose of monitoring charged particles and solar ultraviolet and X-rays. The launching of the first satellite by the Soviet Union on October 4, 1957, with its surprisingly heavy payload, shocked the United States into making an all-out effort to attain leadership in space exploration. A full year went by before the legislative process in Washington created a space agency, but in the interim a number of important initiatives were taken by the military and by private organizations. It is widely assumed that astronomy would be one of the cornerstones of the scientific space program, but until NASA appeared on the scene no one had any sure idea as to how the program would

be organized. The role of industry was particularly unclear and companies were eager to demonstrate their competence in the field. For example, during the first few months of 1958, the McDonnell Aircraft Corporation of St. Louis, under the leadership of their president, James S. McDonnell, developed a detailed plan for the design, construction, launching and operation of astronomical satellites, which they called Astroscopes. After consulting with a number of astronomers, among them R. Tousey, J.W. Evans, L. Spitzer, M. Schwarzschild and myself, they designed a solar satellite or Solarscope[10] which would carry both solar and non-solar experiments. These included a solar ultraviolet spectrograph, a camera to photograph the sun in the light of the Lyman-alpha line of hydrogen, an experiment to measure the intensity and polarization of light from the extra-galactic universe, and a simple radio-astronomy experiment to explore the spectrum below 30 megacycles. In implementing the plans for Solarscope, McDonnell Aircraft proposed a contractual arrangement with the University of Michigan's Department of Astronomy, which I chaired, to assume responsibility for programming and data analysis once the satellite had been placed in orbit.

In parallel with its engineering design studies, the McDonnell company sponsored the University of Michigan to prepare a report on important astronomical experiments that might be carried out with satellites. The report,[1] prepared in the spring and early summer of 1958, was published in November of that year. It was the only document of its kind in existence at the time when NASA came into being and provided valuable scientific input into NASA's early planning for astronomy.

About two-thirds of the McDonnell Report dealt with solar experiments. Virtually all of the proposed experiments were eventually carried out in the Orbiting Solar Observatories and Skylab missions, among them high-resolution spectra to a short

Figure 3. A Spectroheliogram, transmitted from the Solar Ultraviolet Scanning Spectrometer on OSO-IV, shows the Solar Corona in a magnesium-10 spectral line. The dark shaded areas indicate regions where temperatures exceed 1.5 million degrees Kelvin.

ATM EXPERIMENTS PACKAGE

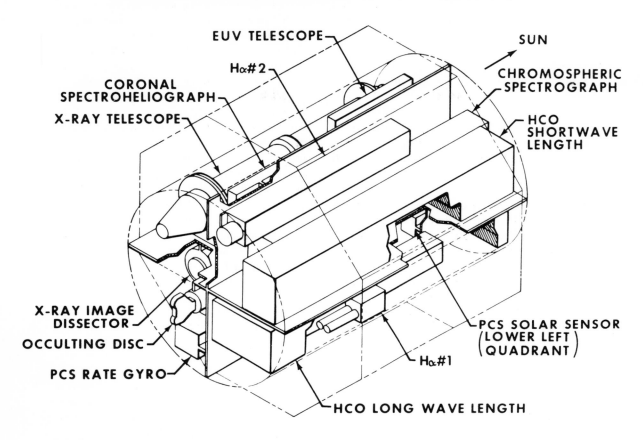

EUV TELESCOPE

Hα#2

CORONAL SPECTROHELIOGRAPH

X-RAY TELESCOPE

SUN

CHROMOSPHERIC SPECTROGRAPH

HCO SHORTWAVE LENGTH

X-RAY IMAGE DISSECTOR

OCCULTING DISC

PCS RATE GYRO

PCS SOLAR SENSOR (LOWER LEFT QUADRANT)

Hα#1

HCO LONG WAVE LENGTH

Figure 4. An early concept of an Apollo Telescope Mount, a stabilized platform to accomodate scientific instruments requiring accurate pointing.

wavelength limit of 1 A, spectroheliograms in hydrogen and helium lines, accurate profiles of the Lyman-alpha line of hydrogen, absolute measurements of spectral energy distribution at wavelengths between 100 A and 3000 A, and measurement of the density of the earth's atmosphere from absorption lines formed along the line of sight from the satellite to the rising and setting sun.

The McDonnell vision for space astronomy, which entailed the design, construction and launching of 10 Solarscopes, together with an associated communication station, was bold and imaginative. But my judgment at the time was that it was quite unrealistic since it implied that NASA would be willing to concentrate astronomical research from satellites at one company and one university. In those early days when development and construction of hardware occurred simultaneously and in an interactive process, it also seemed essential to me

Figure 5. The Skylab space station as photographed in orbit from the Skylab 4 Command and Service Module. The Apollo Telescope Mount is at the rear. Note also the solar shield deployed by the second Skylab crew to shade the Orbital Workshop. Instruments on the ATM gave us our first continuous observations of the Solar Corona over an extended period of time.

that the responsibility for scientific instrumentation be put in the hands of the experimenter rather than the manufacturer.

A second and highly important pre-NASA initiative was taken by the National Academy of Sciences, which established a Space Science Board (SSB) in the spring of 1958 with the late L.V. Berkner as Chairman. Under Berkner's dynamic and aggressive leadership, the Board immediately set out to establish itself as the major source of scientific advice on the organization and content of a national space science program. One of Berkner's first actions was to send out telegrams to leading scientists which invited them to submit proposals for first-generation satellite experiments. At the same time, the Board established several sub-committees, each concerned with a different sub-discipline of space science. The proposals received in response to Berkner's telegram were screened and ranked by the sub-committees, and consequently when NASA began operating in October 1958 it already had a well thought-out scientific program of very high quality.

After considering the experiments proposed for astronomy, NASA decided to proceed by designing a large stabilized platform upon which various scientific payloads weighing up to 700 lbs could be mounted. The basic platform soon became known as the Orbiting Astronomical Observatory and five groups were chosen to instrument four tentatively scheduled missions. The first spacecraft, scheduled for launch in 1963, was to be shared by the Smithsonian Astrophysical Observatory and the University of Wisconsin, the former to compile an ultraviolet map of the sky and the latter to carry out broad-band photometry and low-resolution spectroscopy of stars with a relatively small telescope. The second and third OAOs were to carry 30 to 36-inch reflecting telescopes and high resolution spectrometers, to be furnished by the Goddard Space Flight Center and Princeton University, re-

spectively, while the fourth OAO was to be equipped for solar observation by the University of Michigan under my direction. [Editor's note: In 1960 Goldberg went to the Harvard College Observatory where he became Director in 1966.]

I must confess that I found it staggering to contemplate the preparation of a solar payload weighing several hundred pounds at a time when we had only succeeded in orbiting satellites as large as grapefruits, or at most basketballs. But not yet having reached my present level of sadness and wisdom I was inclined to say "damn the torpedoes and full speed ahead." In the spring of 1959 the five experimenters came together with appropriate NASA staff as a formal working group on Orbiting Astronomical Observatories. Four major instruments weighing a total of approximately 330 lbs would constitute the payload for the first solar OAO.[7] Three of the instruments were to be spectrometers covering the wavelength region 3000 A - 75 A and the fourth a spectroheliometer for imaging the sun in the Lyman-alpha line of hydrogen. Other experiments were considered for inclusion as well: for example, a coronagraph, two or three X-ray counters and a relatively lightweight X-ray imaging device, which would bring the total weight to about 500 lbs.

Other experimenters shared my belief that a more modest beginning for solar satellite astronomy would be prudent, notably Dr. John C. Lindsay of Goddard Space Flight Center. Lindsay persuaded NASA management to accept a proposal from the Ball Brothers Research Corporation of Boulder, Colorado, to build a smaller vehicle based upon the Ball Brothers two-axis rocket pointing control, which was designed to carry 70 lbs of "pointed" instruments. (A "pointed" instrument is one which, in the jargon of the science, is pointed at the sun at all times.) This satellite, which was originally designated S-16 and later OSO-1, successfully entered orbit in March 1962 and became the first in the famous series of Orbit-

Figure 6. Launch of an Aerobee 150 sounding rocket. These tower-launched rockets, capable of carrying a 150-pound payload to an altitude of 150 miles, were used in early observations of the sun from above the atmosphere.

23

Figure 7. The High Altitude Observatory
White Light Coronagraph was one of the six
principal instruments on the Apollo
Telescope Mount of Skylab. This photograph
shows not only the sun's Corona, but also Comet
Kohoutek (1973f) as it neared perihelion. The
comet image is the bright streak in the upper
part of the corona on the right.

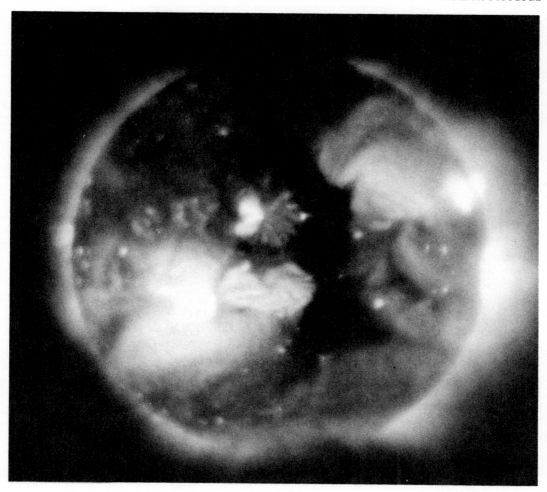

ing Solar Observatories.

The OSO satellites carried instrumentation in two main sections.[7] A slowly spinning wheel, turning at the rate of one revolution every two seconds, maintained the spin axis perpendicular to the direction of the sun within a ten-degree range by gyroscopic action. The wheel consisted of nine pie-shaped compartments, six of which were available for experiments which could look at the sun every two seconds as the wheel rotated and also at non-solar sources that passed through the field of view. The upper section of the satellite carried an array of solar cells, which could be rotated in azimuth about the spin axis independently of the wheel and made to

Figure 8. A large equatorial coronal hole observed at 1110 UT on 1 June 1973 in soft X-rays (2 - 54 A) from the Apollo Telescope Mount on Skylab. This was the first equatorial hole observed from Skylab and played an important part in the identification of coronal holes with the sources of recurrent, high speed, solar wind streams.

face continuously in the direction of the sun. The scientific instruments were carried on the same mounting as the solar cell array, but they were also rotatable in elevation and aimed at the sun with an accuracy of one to two minutes of arc. The upper pointed sections of the OSO satellites carried a variety of instruments, such as ultraviolet and X-ray

25

spectrometers and imaging devices, and a white-light coronagraph. The wheel sections carried experiments for the monitoring of solar and cosmic X-rays and gamma rays in various energy ranges, for the mapping of the zodiacal light, for the detection of neutrons from the sun and for many other purposes.

Although a second working group on Orbiting Solar Observatories formed in 1959, the concept of a larger solar OAO was not abandoned for at least another year, during which it was contemplated that the OSO program would serve as a stepping stone to the OAO. That the technical requirements for solar and stellar satellites were very different soon became obvious and the two programs were separated.

It may sound strange to the present generation of young astronomers, who have always been accustomed to intense competitive bidding for scientific resources, to hear about an entire spacecraft being assigned to a single experimenter or laboratory. But during the period between 1959 and 1964, space astronomy was a very risky undertaking and most astronomers still preferred to take their chances with cloudy weather rather than to suffer the frustrations of operating in space.

Astronomers had good reason to be wary of space undertakings in those early days. To begin with, the proportion of rocket failures was uncomfortably high. Much remained to be learned about operating equipment in the space environment, and problems such as the lubrication of moving parts in a vacuum, the effects of radiation on optical surfaces, and high-voltage breakdown were still largely unsolved. Then there were the unexpected and sometimes terrible accidents that plagued the space program. In 1964, OSO-B was destroyed and three lives were lost in the hangar at Cape Canaveral when the third-stage rocket upon which the satellite had been mounted for spin stabilization was accidentally ignited by the discharge of static electricity from a plastic protective shroud over the satellite. The refurbished prototype of OSO-B was successfully orbited one year later, but the pointed experiments suffered from severe electronic malfunction caused by high voltage arcing. A third OSO failed to reach orbit, which was the kind way of saying it fell into the Atlantic.

My research group at Harvard suffered penalties from the first two of these three disasters, and it was seven years before we achieved our first successful satellite mission with OSO IV in October 1967. It was beginning to look as if the primary purpose of space research was to build character. But as trying as these setbacks were at the time, they made us appreciate the successes all the more when they finally came.

All of these mishaps occurred very early in the program. The last five spacecraft in the series, OSO IV through VIII, all performed according to the best hopes and expectations of NASA and the experimenters. Overall, the OSO series was a sparkling success; under the pressure of scientific necessity, each spacecraft in the series achieved major gains in capability over its predecessors. Thus, OSO I had no imaging capability, but beginning with OSO II, the spatial resolution of the experiments in the pointed section progressively improved, reaching one second of arc in OSO VIII. Raster scanning of the entire solar image was introduced with OSO IV and was further refined in OSO VI, which could be made to point with an accuracy of one minute of arc to any of over 16,000 points in a grid and to perform raster scans covering an area of about seven minutes square around any of these points. (Raster scanning is the TV style of line-by-line reproduction of an image.) We had requested raster scanning as a means of generating monochromatic images of the sun which we had hoped could be made in perhaps a dozen wavelengths, particularly in the Lyman-alpha wavelength of hydrogen at 1216 A and in the resonance-line wavelengths of neutral and ionized helium at 584 A and 304 A, respectively. As

Figure 9. The Solar Maximum Mission spacecraft (launched in 1980) carrying instruments for observation of solar flares in the ultraviolet, X-ray and gamma-ray regions of the spectrum.

it turned out, NASA engineers developed a new and highly sophisticated command system which was capable of pre-selecting any wavelength, at intervals of 0.1 A, within the operating range of the spectrometer, which was 300-1400 A.

As early as 1962, during the Iowa Summer Study[12] sponsored by the Space Science Board of the National Academy of Sciences, astronomers recommended the development of an Advanced OSO, or AOSO, as a follow-on to the OSO. The proposed AOSO, as it was finally approved, would have pro-vided for optics ten feet long and up to 22 inches in diameter with a pointing accuracy of five arc seconds and a pointing stability of one arc second for five minutes. It was intended that the spacecraft be launched by a Thor-Agena B rocket and placed in a polar orbit. The proposed launch date was some-time in 1966. AOSO was cancelled in De-cember 1965, ostensibly for budgetary rea-sons; but there were also doubts that the Thor rocket could do the job. The alterna-tive of using a manned spacecraft began to look attractive.

By the time of the Iowa Summer Study, NASA had already decided that rendezvous and docking in lunar orbit would be em-ployed in the course of returning astronauts from the surface of the moon. If this tech-

HAO SMM CORONAGRAPH/POLARIMETER
DOY 103   UT= 1416 POL=0

Figure 10. A portion of the solar corona from data supplied by the coronagraph/polarimeter on the Solar Maximum Mission Satellite. Note the long coronal spikes, extending from the densest coronal region out to more than a million miles from the sun's photosphere.

nique could be mastered, it would obviously create fantastic opportunities for the assembly of very large structures in space, including astronomical telescopes. When the next summer study sponsored by the Space Science Board convened at Woods Hole in 1965,[15] NASA strongly encouraged astronomers to think of large projects that would require the use of man in space. The space agency expressed strong interest in identifying projects that would test man's capacity to perform scientific tasks in space. In that connection, NASA had begun to organize a program designated as the Apollo Extension Systems (AES) in which Apollo hardware remaining from the lunar landings would be flown for scientific purposes in earth-orbital missions of one to two months.[6] John Lindsay, now head of the Solar Physics Laboratory of the Goddard Space Flight Center, encouraged the Ball Brothers Research Corporation to propose a solar observing facility for the AES program. With a main element called the Apollo Telescope Orientation Mount (ATOM), the plan envisaged a number of AOSO-type experiments mounted on a spar that would be erected from the service module of the Apollo spacecraft and operated by astronauts during prolonged periods of observation from earth orbit.

Later, the AES program was cancelled in favor of a more ambitious undertaking that took a larger step in the direction of a space station. The original plan for this undertaking called for a crew of astronauts to go into earth orbit to recover the expended third stage of a Saturn V rocket and to equip it as a workshop-laboratory. The solar observatory, or Apollo Telescope Mount (ATM), mounted on the ascent stage of the Lunar Excursion Module (LEM) would then be brought up and attached to the workshop through a multiple docking adaptor. Later, this "wet workshop" concept gave way to the concept of a fully assembled "dry workshop" and attached ATM, which would be launched on a Saturn V rocket. Originally scheduled to be launched in 1969 at the time of solar maximum, ATM, or Skylab as it came to be known, did not leave the ground until February 1973. But the added knowledge and experience gained during the extra three to four years of preparation made it possible to extend the length of the observing periods from 28 days to 167 days. It also allowed time to develop a carefully thought-out observing program and a detailed coordination of effort involving the astronomers, the astronauts, and their supporting staff of engineers.

From the viewpoint of a solar astronomer, the first 15 years of the US satellite program were extraordinarily fruitful. I think of the unravelling of the complex structure and dynamics of the sun's outer atmosphere—the chromosphere, the corona and the solar wind; of the discovery and detailed investigation of the coronal holes, of their magnetic origin and their role as the probable source of the solar wind and of the charged particles that trigger off geomagnetic disturbances; of the knowledge gained about the role of magnetic fields generally in heating the corona and shaping its structure; of the exotic high-energy phenomena revealed on the solar surface by X-rays and gamma rays, the further study of which will at last clarify the nature and origin of solar flares; and of the mountain of new information that is still being digested.

The success of the program has been due not only to the scientists who conceived the experiments, analyzed the data, published the papers and devoted large amounts of their scientific lives to the enterprise, but also to the many hundreds of engineers, technicians, computer programmers, managers and secretaries, both in their own laboratories and in NASA and industry who designed and built the spacecraft and experiments, wrote the computer programs, operated the ground stations and otherwise contributed to the massive exercise in cooperation that characterizes every scientific space mission. The enthusiasm and dedication with which these men and women tackled the tough technical problems that we astronomers set before them is one of my warmest remembrances of the early days of the space program.

REFERENCES

1. Aller, L.H., Goldberg, L., Haddock, F.T., Liller, W., *Astronomical Experiments Proposed for Earth Satellites,* University of Michigan Research Institute UMRI Project 2783. November, 1958.

2. Baum, W., Johnson, F., Oberly, J., Rockwood, C., Strain, C., Tousey, R., *Phys. Rev., 70,* 781, 1946.

3. Dobson, G.M.B., *et. al., Proc. Roy. Soc.* Ser. A, *122,* 1929; *129,* 1930.

4. Edlen, B., *Zeitschrift für Astrophysik, 22,* 30, 1943.

5. Friedman, H., *Annual Review of Astronomy and Astrophysics, 1,* 59, 1963.

6. Goldberg, L., *Astrophys. J., 191,* 1, 1974.

7. Goldberg, L.*Proceedings of the 10th International Colloquium on Astrophysics,* Mem. Soc. Roy. Sci. Liege*IV,* 30, 1961.

8. Lyman, Project RAND, Appendix V. Douglas Aircraft Company, Sept. 1, 1946.

9. Menzel, D.H., Harvard College Observatory Circular, 410, 1935.

10. McDonnell Aircraft Corporation Research Department Report Nos. 6309-6313, incl. *Solarscope.* August, 1958.

11. Regener, E. and V., *Phys. Zeits., 35,* 788, 1935.

12. *A Review of Space Research.* (National Academy of Sciences—National Research Council Publ. 1079: Washington, D.C., 1962).

13. Rosseland, S., *Nature, 123,* 207, 761, 1929.

14. Saha, N.M., Harvard College Observatory Bulletin 905, p. 1, 1937.

15. *Space Research, Directions for the Future.* (Space Science Board, National Academy of Sciences—National Research Council: Washington, D.C. December, 1965).

ROCKET ASTRONOMY— AN OVERVIEW

HERBERT FRIEDMAN

Prior to World War II attempts to probe the upper atmosphere directly were hardly very advanced. Robert H. Goddard included atmospheric sensors in some of his pioneering rockets, but the capabilities were not significantly superior to those of meteorological balloons. One important early scientific achievement stands out—the Austrian Victor Hess's discovery of cosmic rays with a balloon-borne ionization chamber carried to a height of 5 km in 1911.[1] But prospects for high altitude research suddenly opened when World War II ended in Europe. V-2's had been produced in the largest underground factory in the world in Nordhausen, Germany. Although the site was within the agreed-upon Russian perimeter, US Army forces got there first and pulled out enough V-2 parts to later assemble about 100 rockets at the White

Herbert Friedman is Chief Scientist Emeritus of the E.O. Hulburt Center for Space Research at the Naval Research Laboratory, Washington, D.C. and is a pioneer in the development of high-energy astronomy.

Sands Missile Range (WSMR) in New Mexico.

When the news spread, late in 1945, that captured V-2 rockets would be brought to WSMR, an organization of scientific users came into being. While the Army studied the engineering details and flight performance of the rocket, its 2,000-pound warhead space was offered for scientific instrumentation. At the Naval Research Laboratory (NRL) the Rocket Sonde Branch was established under the direction of Ernest Krause. He and his staff began preparations of scientific payloads for atmospheric, ionospheric and cosmic ray research. In the Optics Division led by Edward Olson Hulburt there was much enthusiasm for the use of the rockets to study the sun's ionizing radiation, ultraviolet and X-rays. Hulburt, a pioneer in ionospheric research, had participated in auroral research in the second International Polar Year, 1932-33. It occurred to him that it might be possible to photograph the solar ultraviolet spectrum with the same quartz spectrograph that had been used in the auroral studies a dozen years earlier. His proposal was taken up by Richard Tousey and his colleagues, but they quickly realized that Hulburt's little spectrograph was not likely to succeed. Within three months an innovative design developed into flight hardware and was prepared for launch in mid-1946.

Prior to the beginning of high altitude research with V-2's the solar spectrum could be observed from the ground only to wavelengths about as short as 2900 A (angstroms) in the ultraviolet, where atmospheric ozone becomes opaque. The Explorer I and II balloon flights of 1934 and 1935, rising to 80,000 ft over Germany, failed to show any ultraviolet extension of the sun's spectrum.[2] Attempts to observe in an atmospheric "window" between the absorption bands of ozone (O_3) and molecular oxygen (O_2) at about 2150 A from a height of 3500 meters on the Jungfrau mountain were also inconclusive.[3]

The Solar Ultraviolet

The first NRL attempt at solar spectroscopy, which was launched on a V-2 on June 28, 1946, ended in catastrophic failure. Returning to earth nose down in streamlined flight the rocket created an enormous crater some 80 ft. in diameter and 30 ft. deep. Only a bucket-full of identifiable debris was recovered, as though the rocket had vaporized completely on impact. On the next attempt, October 10, 1946, the spectrograph was mounted in the tail section which separated from the rest of the rocket before reentry. It fell relatively gently to earth and the film was recovered undamaged. The spectrograms showed that ozone was concentrated in a thin stratospheric layer centered at a height of about 25 kilometers. As the ozone was penetrated the solar spectrum revealed itself deeper and deeper into the ultraviolet region.[4]

J.J. Hopfield and H.E. Klearman of the Johns Hopkins University Applied Physics Laboratory followed the early NRL success some six months later with a similar achievement.[5] But as observations continued to improve, it became apparent that more sophisticated instrumentation would be needed. In particular, it was essential to have a stablized platform to permit continuous pointing at the sun while the rocket spun, precessed and yawed. The effective exposure could then be increased by a large factor during the few minutes of rocket flight and the spectrum could be followed much further into the ultraviolet where the intensity of radiation is very weak. Several years passed, however, before a biaxial pointing control was developed at the University of Colorado under contract to the Air Force Cambridge Research Laboratories.[6] Thereafter persistent efforts pushed the limit of the observed spectrum far into the ultraviolet and revealed that the overall spectral intensity fell rapidly from the 6,000° Kelvin blackbody, which characterizes the visible region, to about 5,000° K near 2,000 A and to 4500° K near 1500 A. At X-ray wavelengths it seemed clear that the amount of radiation from the photospheric disk of the sun that is most luminous in the visible band would be utterly negligible.

Solar Control of the Ionosphere

Marconi's early success in transmitting radio signals across the Atlantic confirmed the existence of a reflecting layer of the earth's atmosphere, the ionosphere, above a height of about 100 km. Reflection requires a "mirror" of free electrons. Each electron oscillates in tune to the radiowave frequency that it picks up and, like a small radio-relay station, rebroadcasts the signal to distant areas on the surface of the earth. The existence of these electrons in a free state, torn from the normal atmospheric gases, requires short wavelength ultraviolet rays or X-rays.

Although the diurnal growth and decay of the ionosphere clearly reveals the direct con-

Figure 1. Six Nike-Asp rockets mounted on rail launchers aboard USS Point Defiance for eclipse study in South Pacific, 1958.

trol of the sun, radio scientists were baffled to explain how a 6,000° sun could produce the required electrification. Hulburt recognized that soft X-rays had the right penetration to deliver their ionizing (electron freeing) energy in the region of the atmosphere from 90 to 130 km called the E-region. But how could the sun produce X-rays? The most likely source had to be the solar corona, a great aura of faint, pearly light that envelopes the disk. The corona became visible only when the brilliant orb is masked by the moon in an eclipse. Assuming that the corona was bound by solar gravity, its great extent required a very high temperature, perhaps millions of degrees, to support it against the pull of gravity. Million-degree plasma must radiate X-rays at wavelengths of a few angstroms, the range that would be most conducive to ionizing in the ionospheric E-region. But how intense was the X-ray emission? Efforts to obtain the answer were first made with film packets exposed on the surface of a V-2 rocket.[7] Although film blackening—that is, a very high intensity—was registered in early flights, the results were somewhat suspect.

Figure 2. First solar spectrum obtained at high altitude, 10 October 1946. (Baum et al, 1946).

First Generation X-Ray and Ultraviolet Photometry

In 1949, V-2 No. 49 carried Geiger counters sensitive to a narrow band of X-rays centered about 8 A. As the rocket climbed to a peak altitude of 150 km, X-rays were detected near 85 km and increased in intensity to 120 km.[8] The observed flux was consistent with the required photoionization rate of the ionospheric E-region by thermal radiation. The same rocket carried detectors for Lyman-alpha emission (1216 A), the resonance line of hydrogen, and for the Schumann continuum region of the ultraviolet region (1425-1600 A). Lyman-alpha, which originates in the hot solar chromosphere a few thousand km above the photosphere of the sun, contained most of the energy in the solar extreme ultraviolet and was absorbed between 75 and 90 km above earth, in the ionospheric D-region. We understood only later that Lyman-alpha controls the D-region when a mass spectrometer flown through the ionosphere revealed the presence of nitric oxide. Although only a trace constituent at a concentration of 10^{-8} par/cc nitric oxide is so efficiently ionized by Lyman-alpha that it provides nearly all the normal ionization of the D-region. The radiation detected in the Schumann region produces no ionization

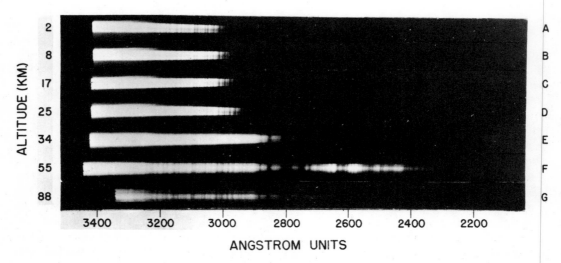

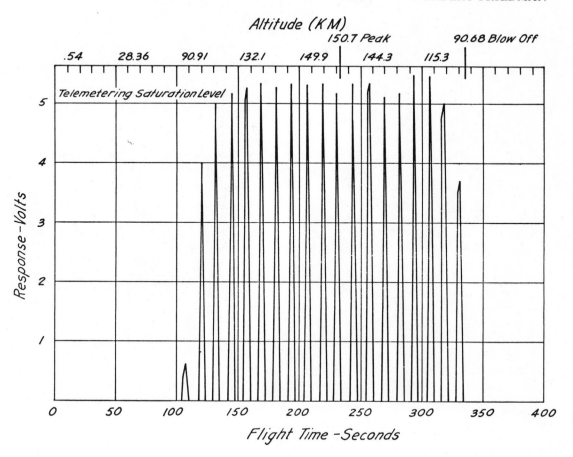

Figure 3. Telemetry record of solar X-ray flux (1-8A) obtained from spininng V-2 rocket, 29 September 1949. (Friedman et al., 1951).

but plays a very important role in shaping the ionosphere by dissociating molecular oxygen. O_2 controls the rate of neutralization of ionospheric electron density much more effectively than O.[9]

Early Rocket Experiences

The 2000 lb warhead space of the V-2 was in some respects an embarrassment to the scientists involved in early flights. Individual instruments were typically simple and even though miniaturization was not yet highly developed the payloads were generally underweight. Concrete or lead ballast was frequently added to maintain the proper distribution of mass for stable flight. Payload space was often parceled out to several experimenters, and instruments were inter-

faced with insufficient testing. Odd assortments of packets of frogs eggs, seedlings, and cosmic ray emulsion packs were often taped into whatever free space could be found just before flight. Inevitably, failures occurred from unanticipated interference between different experiments. Some incompatibilities bordered on the absurd. Emulsion packages, for example, traveled by air mail from the Air Force Cambridge Research Labs to be installed by staff at WSMR. The presence of the film remained unknown to NRL staff who tested their X-ray Geiger counters in place in the rocket with

5 mg. radium sources. When developed after flight, the emulsions contained an enormously high density of electron tracks. The Air Force scientists were baffled; the Navy scientists were innocently ignorant of the damage they were inflicting on the emulsions.

V-2 rockets were a free gift to scientists, but their performance was agonizingly unpredictable. Some exploded on ignition; some somersaulted end over end; one landed on the edge of Juarez, just over the Mexican border. There was clearly a need for a rocket more suitably tailored for scientific research. The Aerobee quickly became the research rocket. James Van Allen had specified its basic characteristics while he was at the Johns Hopkins University Applied Physics Laboratory. In its first appearance it presented several problems. Starting with a severe jolt and a 15-g accerleration, it barely reached 90 km. Several improvements quickly made it more attractive for research. After stretching its tank section a couple of feet, its maximum altitude exceeded that of the V-2 and it became the favorite of researchers for the next decade.

NRL undertook the development of the Viking rocket to provide several fundamental improvements over the V-2. It had a comparable payload-weight capability plus spin and attitude stabilization. But its development turned out to be better timed for its use in the Vanguard program than as a routine research vehicle for the pre-Sputnik era.

Solar Cycle X-Ray Variability

Although the sun appears to be very stable in its output of visible light, there are many signs of variable activity when it is observed with telescopes that are equipped with sophisticated filters and spectrographic instruments. Most prominent are the sunspots, which come and go with an ap-

Figure 4. Early V-2 prepared for launch at White Sands.

Figure 5. Crater left by impact of vanished V-2 at White Sands.

proximate 11-year periodicity.

Following the 1949 measurement of hard X-rays (1-8 A), broadband photometry of the X-ray spectrum was progressively improved with the use of various filters and detector window materials—beryllium, aluminum, titanium, mylar, formvar, etc. The X-ray spectral distribution could be identified with thermal radiation from a corona at a temperature of a few million degrees. Successive measurements at time intervals of months to years showed strong evidence of flux variations of as much as a factor of 7 for X-rays (8-20 A) over the span of the solar sunspot cycle. Such behavior was consistent with ionospheric electron density trends in the E-region, and it appeared that the solar X-ray flux controlled most of the E-region.[10]

The X-Ray Corona

X-ray flares and slower variations in solar X-ray flux made it evident that the concept of X-ray emission from a spherically symmetrical hot corona was altogether inadequate. The corona had to be structured so that densifications formed over sunspots and thus produced enhanced localized X-ray emission. Since sunspots are characterized by powerful magnetic fields, the condensations of the corona were suspected to form in tight loops of magnetic lines of force, spanning the gap between adjacent spots of north and south polarity. To confirm this model posed a difficult challenge because no X-ray telescope was available in the early years to image the X-ray sun, even though in principle such a reflecting telescope could be constructed.

During the International Geophysical Year (IGY), 1957-58, the availability of two-staged solid propellent rockets made it feasible to plan a series of rapid firings in the course of a solar eclipse. If X-rays came from a few discrete active regions overlying sunspots, the detectors would register rather abrupt decreases in the X-ray flux as the moon occulted each source region. The eclipse of October 10, 1958, tracked across the South Pacific, missing all ground sites except for a tiny coral atol named Puka Puka in the Danger Islands. To support a team of ground based astronomers, as well as a series of rocket firings, the Navy provided a Landing Ship Dock, the USS Point Defiance. Six Nike-Asp rockets were mounted on its

Figure 6. Rockoon rising from deck of USS Colonial off the coast of California.

helicopter deck, while the ground-based equipment was delivered ashore. At eclipse time the rockets were launched at appropriate intervals to observe the moon's passage across the major sunspot groups on the solar disk. The telemetry records showed that X-rays came from high in the corona. Even with the disk completely hidden at totality, 13% of the X-ray flux persisted. As the moon cut across the individual sunspot groups the X-ray intensity dropped abruptly. The expedition thus succeeded in proving the connection of X-ray emission with coronal condensations high above sunspots.[11]

From 1958 on, substantial efforts were made to perfect imaging X-ray telescopes, but the first solar X-ray photograph was obtained with the most elementary means. In 1960 a simple pinhole camera rode a pointing control aboard an Aerobee rocket. It registered an X-ray image showing 80% of the emission concentrated in only 20% of the area of the disk.[12] From that modest beginning solar X-ray astronomy moved on to the use of reflecting telescopes and the achievement of images with a few arc seconds of resolution.

Solar Flares

Solar flares are the most explosive manifestations of solar activity, releasing energy comparable to that of billions of hydrogen bombs in less than an hour. Their occurrence is not yet predictable except that the number of flares increases approximately with the number of sunspots, and only the release of the enormous energy stored in sunspot magnetic fields can match the energy production of a flare.

Since the beginning of radio communication via the ionosphere, it has been known that large flares produce radio blackout or fadeout. Before the discovery of solar X-rays, theorists tried in vain to explain the radio blackout in terms of a flash of ultraviolet radiation, but the energy required was many orders of magnitude greater than

Figure 7. Nose cone and instrumentation section for attachment to Nike-Deacon on rail launcher in background.

was theoretically possible. On the other hand, a comparatively modest burst of X-rays at the shorter wavelengths could theoretically create the condition for fadeout.[13] It may seem paradoxical that soft X-rays produce a reflecting ionosphere, but hard X-rays produce a reflection blackout. The explanation lies in the level of absorption of hard X-rays at 60 to 80 km above the earth. At these lower heights and much

higher ambient densities, electrons that are ripped from atoms by X-rays do not remain free long enough to reflect radiowaves. They collide very quickly with gas molecules, and the energy which is picked up from incident radio waves is dissipated as heat. Only a millionth part of the solar flux in the form of 10 keV solar flare X-rays is sufficient to cause radio fadeout. To test this model of X-ray-induced radio blackout a major effort was planned for the IGY.

Because flares give no precursor signals, it was impractical to set up an instrumented rocket at White Sands to be launched at the moment of a flare outburst. V-2's, Vikings and Aerobees, which made up the stable of research rockets at the time, all used liquid propellants that could not be stored in the rocket while it stood on the launch stand for more than a few hours. A military rocket, the Deacon, nine feet long and six inches in diameter, was the only solid-propellant vehicle available, and it could reach only 40 km when launched from the ground. If the Deacon were carried to 25 km on a balloon, however, it could be ignited at that altitude and would then climb to well above 100 km. This combination of rocket and balloon was named the "Rockoon" by its inventor, James A. Van Allen.

In the pre-IGY Year, 1956, an expedition called "Operation San Diego High" was organized with support from the U S Navy.[14] That summer, a Landing Ship Dock, the USS Colonial, moved out to sea about 400 miles off the coast of southern California, with ten Rockoons. In the early morning of each day a 150,000 cubic foot polyethylene balloon carried a Deacon aloft. As the Roc-

Figure 8. Nike-Deacon leaving rail launcher for solar flare study at San Nicholas Island during IGY.

Figure 9. X-ray (0-60 A) image of the sun, 19 April 1960 obtained with pinhole camera aboard Aerobee rocket. Rotation of camera about pointing direction during exposure smeared discrete emission regions into arcs. (Blake et al, 1963).

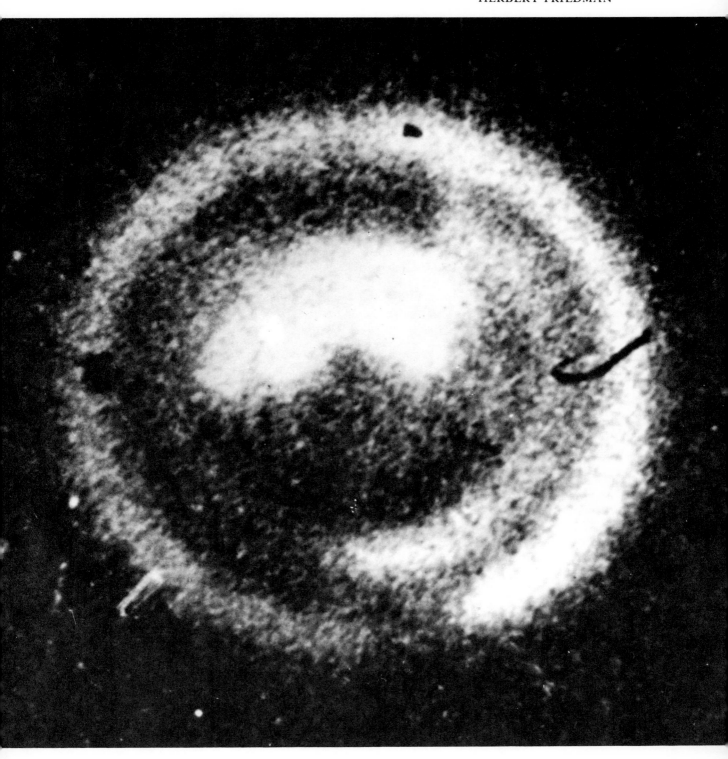

koon floated at 80,000 feet scientists aboard the LSD could communicate by teletype with the High Altitude Observatory at Boulder, Colorado, and with the Sacramento Peak Observatory in New Mexico. When the rocket team was alerted that a flare had started, the Rockoon could be fired by radio command. It would then streak above the atmosphere to measure the ultraviolet and X-ray flux as the flare reached the peak of its flash.

On the first three days the rocket floated all day, but no flare appeared on the sun. At the end of each day the rocket had to be fired to avoid any possibility of it drifting back over land. Success was achieved with the fourth rocket. The instruments detected X-rays from a modest flare, sufficient to account for the radio fadeout. At the same time the ultraviolet intensity remained essentially unchanged, proving that it contributed negligibly to the ionospheric absorption.

Beginning about 1957, two-stage combinations of solid-propellant rockets such as the Nike-Deacon replaced the Rockoon for flare studies. A substantial number of these rockets were committed to the IGY solar flare studies from 1957 to 1959. Launches took place on San Nicholas Island just a short distance off the California coast. A substantial number of large flares were observed. All showed X-ray flashes extending well above 20 keV in energy.[15] In the strongest flares, X-rays penetrated to altitudes as low as 43 km. The spectra of the flares were typical of super-heated plasma: temperatures reached a hundred million degrees. By the end of the IGY, these pioneering rocket measurements were replaced by satellites dedicated to solar observations, which provided an almost continuous flare watch on the sun, more effective than hundreds of rockets each providing only a few minutes of observations.

Airglow and Stars in the Ultraviolet

From the very beginning of upper air re-

search with V-2's, measurements of the basic parameters of the atmosphere—pressure, temperature and density—were given high priority. Once solar X-ray and ultraviolet measurements became standardized, they provided very precise indirect density and composition measurements. High altitude airglow photometers for the oxygen red and green lines showed that ground-based estimates were badly in error. But perhaps the most surprising atmospheric discovery was the vast cloud of hydrogen surrounding the earth—a geocorona 50,000 miles high. In 1955 an Aerobee rocket fitted with detectors for Lyman-alpha, the hydrogen ultraviolet resonance line, found the entire night sky above 85 km suffused with its glow. Theoretical analysis showed that the solar Lyman-alpha radiation was being scattered back to the dark side of the earth by the extended cloud of geocoronal hydrogen.

A more elaborate experiment carried out in 1957 with an Aerobee rocket fitted each ultraviolet photon counter detector with a collimator composed of a bundle of hypodermic needles. The view through the collimator had an angular field of 3 degrees. As the rocket spun and yawed it allowed the detectors to produce a spatial scan of much of the sky. Not only was the shape of the geocorona more precisely defined, but numerous stellar sources were detected. Early stars stood out brightly in the extreme ultraviolet.[16]

The first flight of a reflecting telescope for stellar ultraviolet astronomy took place in November 1959. The telescope consisted of a four-inch parabolic mirror which focussed an image of a point source on an ultraviolet ion chamber. This simple instrumentation was sufficiently sensitive to provide an early catalogue of about 200 of the brightest early type stars.[17] By 1960, NASA had committed a major effort, the Orbiting Astronomical Observatories, to ultraviolet astronomy.

X-Ray Astronomy

Success in solar X-ray observations natu-

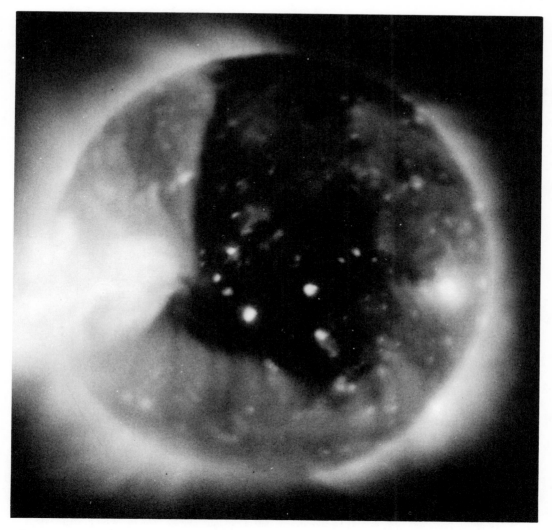

rally led to speculation about the possibility of galactic X-ray astronomy. If the only potential X-ray sources were stars like the sun, the prospect of detection seemed hopeless. From the distances of even the nearest stars, the flux at earth would be many orders of magnitude below the sensitivities of the instruments of the 1950s era. But radio astronomers had observed extremely powerful sources in the galaxy as well as distant radio galaxies. Their mechanism of radiation was synchrotron emission generated by very high-energy electrons circulating in magnetic fields. It seemed within the realm

Figure 10. The largest equatorial coronal hole ever observed in soft X-rays (8 - 64 A) is seen in this photograph taken from a sounding rocket at 1948 UT on 27 June 1974. A connection between the equatorial coronal hole and the polar hole is visible in the northeast.

of reasonable speculation to expect the radio spectra to continue through the visible and ultraviolet, all the way to X-rays.

A stimulus for the search for galactic X-ray sources came as a by-product of the solar flare Rockoon expedition. Each rocket was equipped with a scintillation counter for hard X-rays/gamma-rays in addition to a complement of thin-window X-ray proportional counters. On several flights, hard X-ray signals were observed at lower altitudes, but they softened as the rocket climbed through the atmosphere. A clear residual X-ray flux below 40 keV seemed to indicate an extraterrestrial source.[18] To eliminate positively the possibility that the radiation was from the sun required that the observations be repeated at night, but the crucial tests were not accomplished by the NRL group in subsequent years. In 1962, positive evidence for galactic X-rays was finally obtained by scientists of MIT and American Scientific and Engineering. From that point on, the new X-ray astronomy has grown rapidly to the present Einstein Observatory. The X-ray sky now offers neutron stars, black holes, white dwarfs, supernova remnants, Seyfert galaxies, BL-Lac objects, globular star clusters, galaxy clusters and superclusters and even ordinary stars of all classes from early to late type in the main sequence.

NOTES

1. V.F. Hess, *Phys. Zeits., 13,* 1084, 1912.

2. E. Regener and V. Regener, *Phys. Zeits., 35,* 788, 1934.

3. E. Meyer, M. Schein and B. Stoll, *Helv. Phys. Acta, 7,* 670, 1934.

4. W.A. Baum, F.S. Johnson, J.J. Oberly, C.C. Rockwood, C.V. Strain and R. Tousey, *Phys. Rev., 70,* 81, 1946.

5. J.J. Hopfield and H.E. Clearman, *J. Opt. Soc. Am., 37,* 405, 1948.

6. D.S. Stacey, G.A. Smith, R.A. Nidey and W.A. Pietenpol, *Electronics, 27,* 149, 1954.

7. T.R. Burnight, "Physics and Medicine of the Upper Atmosphere," p. 226, University of New Mexico Press, 1952.

8. H. Friedman, S.W. Lichtman and E.T. Byram, *Phys. Rev., 83,* 1025, 1951.

9. R.J. Havens, H. Friedman and E.O. Hulburt, "The Physics of the Ionosphere," Rep. Phys. Soc. Conf., p. 237, 1955.

10. H. Friedman, Proc. Inst. of Phys. and The Physical Society, Conf. on "The Ionosphere," London, July, 1962, pp. 3-18, 1963.

11. T.A. Chubb, H. Friedman, R.W. Kreplin, R.L. Blake and A.E. Unzicker, Proc. 10th Intl. Astrophys. Symp., Liege; *Mem. Soc. Roy. Sci. de Liege, 20,* 228, 1961.

12. R.L. Blake, T.A. Chubb, H. Friedman and A.E. Unzicker, *Astrophys. J., 137,* 3, 1963.

13. H. Friedman and T.A. Chubb, "The Physics of the Ionosphere," Rep. Phys. Soc. Conf., p. 58, 1955.

14. T.A. Chubb, H. Friedman, R.W. Kreplin and J.E. Kupperian, Jr., *Nature, 179,* 861, 1957.

15. T.A. Chubb, H. Friedman and R.W. Kreplin, *J. Geophys. Res., 65,* 1831, 1960.

16. J.E. Kupperian, Jr., E.T. Byram, T.A. Chubb and H. Friedman, *Planetary and Space Science, 1,* 3, 1959.

17. E.T. Byram, T.A. Chubb and H. Friedman, *Space Research,* Proc. of 1st Intl. Space Science Symp., Nice, p. 599, 1960.

18. H. Friedman, AGARD Symposium, Paris, France, 1959, AGARDograph 42, pp. 3-10;
    J.E. Kupperian, Jr., and H. Friedman, "Gamma Ray Intensities at High Altitudes," Fifth CSAGI Assembly, Moscow, U.S.S.R., 1958.

EXPLORING THE MOON

ROBERT JASTROW

I joined NASA at the time of its formation in November 1958. Homer Newell, who had been a Division Head at the Naval Research Laboratory, where I was a nuclear physicist, liked the work I had done on the orbit of the Sputnik rocket, and invited me to come into the new Space Agency with him.

My job was to set up a Theoretical Division to conduct research in theoretical physics related to the space program. My little Division was one of three research units; the others were the Vanguard Division under John Hagen and the Space Sciences Division under John Townsend. Later these three units were supposed to be incorporated into the structure of the Goddard Space Flight Center, but the Center did not exist yet. In the interim, we were attached to NASA Headquarters and reported to Keith Glen-

Robert Jastrow is the founder and director of NASA's Goddard Institute for Space Science. He is professor of astronomy and geology at Columbia University, and adjunct professor of earth science at Dartmouth College.

nan, the head of NASA, through Newell and Dr. Abe Silverstein.

The new Division was located in the old NACA building at 1512 H Street in Washington, next to the Dolley Madison House on the corner, which had formerly been the home of the Cosmos Club. The Administrator of NASA and his staff occupied the Dolley Madison House, while the next level of management, and the vineyard workers, such as myself, were quartered in the NACA building.

The offices of the Theoretical Division consisted of three or four rooms in the rear of the sixth floor, into which some ten or twelve mathematicians and physicists were jammed. Homer and his staff had a little more room up front on the same floor.

My first major hire was Dr. John O'Keefe. He agreed to come into our new outfit as Assistant Chief of the Theoretical Division. John promptly made a nice discovery, with Ann Eckels, of the "pear-shaped earth." His work was the Division's first important research result.

Meanwhile, I set about learning what we

needed to know to carry out our mission in the Agency. The mission was to conduct research in all fields of physical science that could be investigated with data acquired from spacecraft. This meant astronomy, the moon and planets, and the earth's atmosphere. These fields were new to me because I had always worked in quantum mechanics and nuclear physics—the worlds of the small in contrast to the astronomer's world of the large. Where to begin?

I came across a book by Harold Urey, the great American man of science, called *The Planets*. Instead of being a dry discussion of the solar system, as such books usually are, it was enlivened by a sense of evolution in the Cosmos and the place of our planet in the larger scheme of things. I detected that cosmological spark, and I thought, "This is the approach to the subject that I must learn, and Harold Urey is the person I must learn it from."

Meanwhile—this was November 1958—I had been talking to John O'Keefe about coming over to NASA, and John had a personal acquaintance with Urey, whom I had never met. He picked up the phone in my office and called up Harold to introduce me. We agreed on a date, and I flew out to Los Angeles on December 3rd and drove down to La Jolla, where Harold was teaching on the new campus of the University of California.

We met the next afternoon in Harold's office. He sat me down, gave me his book on the planets, and opened it to the second chapter on the moon. I had looked through the book before, but not with understanding. Now Harold explained to me how the distorted shape of the moon—out of round, and with a kind of nose or frozen tidal bulge, pointed toward the earth—proved that the moon was cold and rigid. If it had been warm and partly molten inside, like the earth, the frozen tidal bulge would have collapsed into the soft interior.

If the moon was cold, that meant it was geologically dead; therefore, no mountain-building, no lava, and no volcanoes. Therefore, all the craters on the moon must be made by meteor collisions and not volcanic calderas, said Harold, coming down cleanly on one side of the old lunar crater argument. True, there were frozen lava lakes on the surface, but they must be a remnant of some ancient volcanoes that had long since died out.

This kind of deductive reasoning, in which you take a basic principle—the fact that the moon's irregular shape proved the body of the moon was rigid and therefore cold inside—was very appealing to me, because as a physicist I had been trained in that kind of reasoning, and it still dominates my scientific thinking. From one fact—actually three little facts that gave the shape of the moon—plus the laws of theoretical physics, Harold deduced conclusions that explained all the myriad surface features and minutiae visible on the moon's surface.

This was the beginning of the hot-moon cold-moon battle, which Harold fought single-handedly with the geologists. Geologists tend to follow a line of *inductive* reasoning, in which you collect a large number of pieces of rock, or beetles, or butterflies, as the case may be—and then, laying the samples out on the table in a row, you try to induce from this display of details a few general conclusions. At least, at that time geologists tended to reason inductively. Later, continental drift and plate tectonics entered the field and provided a deductive framework for geology; but continental drift did not become accepted until the late 1960s, nearly ten years after the period we are talking about.

In any case, I was very impressed by the deductive style of Harold's reasoning, which led him to a very important conclusion that dominated his thinking about the moon and dominated my thinking about the matter from that moment on. Harold pointed out that the earth had lost its record of its early years—the magic period when life was evolving here—because volcanic eruptions and

46

erosion by wind and water had covered up the old rocks or washed them into the sea. But the moon, he said, being airless, waterless and geologically dead, was not affected by these forces of change. It still held the record of its birth and early years. It could tell us something we would never learn on the earth; it could help us solve the mysteries of the origin of the solar system and the origin of life.

What a heady thought. And all this from that cold, dead, miserable piece of real estate. I resolved to bring Harold's fascinating message to the people in NASA, who were not terribly interested in the moon at that time; in fact, from a scientific point of view they did not know it existed. They were interested in Project Mercury, in the Van Allen belts, in orbiting telescopes, and, some day perhaps, in landing on Mars and Venus.

Nobody at that time, except Harold and one or two others, knew that the moon held unique treasures from a scientific point of view.

So I invited Harold to come East and give some lectures on the moon and the planets at NASA. He gave the lectures on January 15 and 16, 1959, in the NACA conference room on the 9th floor of our building, to an audience of space-science neophytes and a sprinkling of NASA officials. As I recall, a portrait of the Wright Brothers looked down on the proceedings.

Afterwards we came downstairs and sat in my room on the sixth floor rear, and talked about the fact that the Russians were wiping up the floor with us in space. Harold said, "You know about the plans JPL is working on for a soft landing on the moon in 1963 or 1964—why don't we get on it before then and show the world we can do something scientifically important in space." I said, "That is a great idea," and we went up to the front office where Homer Newell was, and I said to Homer, "Harold has been explaining to us what a scientific treasure the moon is and how it holds the key to the origin of the earth, and we have been talking about the

way the Soviet Union was walking all over us in the eyes of the world, and we wanted to advocate a crash project to land on the moon in 1961 instead of 1963 to show the world what we can do.

Homer listened to us and he was very interested, and he said, with his customary clarity and dispatch, "Write up a memo outlining your reasons and recommendations for me to show to Abe Silverstein." We went into the back room and we wrote up the draft that Homer asked for. I have it in front of me. The scientific part is in Harold Urey's language; I recognize the phraseology. I guess he dictated it because it is a typed draft with changes in his handwriting. I edited the political part; for example, I changed "special program" to "crash project," and so on.

The draft starts with a statement of the national purpose which I wrote out with Harold's help:

"Effort should be concentrated on a single project of great scientific value and impact. Successful and rapid completion of such a project will enhance the reputation of the United States to a degree that cannot be achieved by the execution of a conventional scientific program on a normal schedule. With this purpose in mind, we propose that a crash program be set up for the execution of a lunar soft landing in 1961. The project is presently scheduled for the approximate period 1963-64. We presume that it occupies a similar place in the USSR schedule of space research.

"A soft landing with performance of the experiments listed below will capture the imagination of the scientific community and the general public to a greater degree than any project of comparable scientific value."

The statement continues with the scientific purpose of the lunar landings:

"The problem of the origin of the solar

system is one of extreme importance to the origin of the Universe itself. It is one of the great problems with which the mind of man has been concerned since prehistoric times. Study of the moon's surface is intimately bound up to this problem. In fact, there is written plain to our eyes on the surface of the moon the history of the origin of the solar system.

"It is our opinion that a study of the moon is more important than a study of Venus or Mars."

This was the first statement in the Space Agency of the scientific rationale for exploring the moon. It was also the first time anyone ranked the moon as more important than Venus or Mars.

We took the draft over to Homer again and he liked it. Until then, the thinking in the Space Agency was that the moon was a way station enroute to other things, but Harold convinced me that the moon should be the primary thrust in space exploration. That is the view we communicated to Homer, and it impressed him very much. As far as I know, that moment in Dr. Newell's office was really the beginning of the lunar science program in the Space Agency.

We got started right away. Homer agreed to back me in setting up a meeting at JPL, at which we would draw on the wisdom already accumulated at JPL and incorporate it into a NASA planning project for a lunar soft landing. He wrote a letter to Al Hibbs at JPL saying that the meeting would formulate "design criteria and support needs for lunar exploration," I called Al Hibbs and set up the meeting for February 5, 1959, in California.

John O'Keefe and I went out to the meeting carrying a memo addressed to the "Working Group on Lunar Exploration," saying ". . . the project of lunar surface exploration has been adopted by NASA as a part of its project. It will be the responsibility of the present group to see to the accomplishment of this project." This was pretty strong language, but Homer had au-

thorized me to use it.

When O'Keefe and I arrived we saw the situation was sticky because Al Hibbs had been directing this planning effort as a JPL and West Coast affair, but Homer had directed me to make it plain that it was going to be a NASA project, planned and administered out of Headquarters. I have a vague recollection of a contest of wills in which I had to assert myself with some authority, but the atmosphere was not acrimonious once this point was established, and it was a most fruitful meeting from which we parted with an agreement to meet again in Washington on February 14.

At the February 14 meeting we got down to brass tacks with a discussion of payloads, weights, launch vehicles, etc. The planning pointed to the "instrumented crash landing in twelve months"—i.e., early 1960—as having "major impact both on the scientific community and the general public." This proposal was implemented in the JPL Ranger spacecraft which hit the moon with a TV camera that radioed back high-resolution photos of the lunar surface just before it hit. But our hopes for the "crash project" were not realized. The first successful Ranger mission occurred in 1964.

The February 14 meeting also drew up a list of experiments for the payload of the soft lander, which later became the Surveyor spacecraft. Most of the experiments we proposed were actually flown on Surveyor. But the timetable, which Harold and I had hoped to get pushed up from 1963 to 1961 to counter the Soviet successes, remained where it had been. The first soft landing took place in 1966.

Those two meetings at JPL and NASA Headquarters on February 5th and February 14th laid out the main lines of the unmanned lunar program as it eventually developed. I gave Homer a report on those meetings on February 27, 1959. He considered the matter for a few weeks, discussed it with the people involved in vehicles and engineering, and then acted decisively. In a memorandum to his boss, Abe Silverstein, Director of Space Flight Development, dated March 23, 1959, Homer proposed for the first time that a major project of lunar exploration should be a part of the National Space Program, basing his recommendations on the scientific rationale and payload discussions of the two February meetings.

That was it. Harold Urey was the trigger, I was the bullet, and Homer Newell fired the gun. The moon had entered NASA planning.

I do not remember the details that followed, but Homer's memo must have been approved because in the Administrator's progress report for June, 1959, there is a page entry for "Lunar Exploration Project: Robert Jastrow, Project Officer," referring to "firm decisions on the payloads for the first six lunar missions."

All through this period, Harold Urey and I (and no doubt others as well) were thinking about lunar exploration as the best American response to Soviet space achievements. I sent Homer a memo on April 20, 1959, saying,

"Prof. Nesmeyanov, President of the USSR Academy of Sciences, recently stated in his speech to the Academy: 'There is no doubt that such gigantic tasks as the attainment and exploration of the Moon and subsequently of the nearest planets will be attempted before the current 7-year plan ends.'

"It may be asked whether the Office of Space Flight Development is prepared to accept the possibility that the USSR will carry out experiments on the lunar surface before the NASA program has reached this stage. . . .

"It is suggested that the national space program will be open to strong criticism if a very early and vigorous effort is not made in the program of lunar exploration, including specifically the first lunar orbiter scheduled. . . . The criticism will be especially strong if it turns out that a slow-paced US lunar program must be

contrasted with early Soviet achievements in this field."

As matters turned out, there was a flaw in our reasoning. We did land on the moon with elaborate packages of instruments, starting in 1964—three years later than we had hoped, but still a very impressive achievement—but the feat did not have the impact on world opinion that we had expected. That came only when NASA landed *men* on the moon. Then the US finally made its point. Actually, the astronauts did an enormous amount of excellent science on the moon, but I think the impact of the manned lunar landings on world opinion would have been just as great if the astronauts had taken a few snapshots and then come home again.

Our group, which was now formally called "The Lunar Exploration Working Group," met six times in 1959. It included some brilliant people in the scientific community: Harold Urey, Bruno Rossi, Joshua Lederburg, Gordon MacDonald, Frank Press, Maurice Ewing, Thomas Gold, Harrison Brown, Jim Arnold. As you can imagine, the debates were hot and heavy, to say the least. John O'Keefe added a lot of sparks from the Government side, and I kept things moving as the chairman.

In December 1959 we organized a press conference for members of the Lunar Exploration Committee, at which Harold Urey told the reporters about a conference on the moon we had held the day before in the Theoretical Division. I think this was the first chance the reporters and the public had to hear Harold's ideas about why the moon was a scientifically invaluable treasure. Later on, I backed him up with some remarks on the moon as the Rosetta Stone of the solar system. At that same conference the reporters were also treated to a lively argument between Harold and Tom Gold about how much dust there was on the moon. Tom had been maintaining for some time that there could be such deep pits of dust on the moon that spacecraft would be swallowed up when they landed. Recently in The New York Times Tom said he had been misquoted; he only said there would be a few inches of loose dust on the moon, which is, of course, what we found. But I remember it differently.

Meanwhile, the lunar project continued to gather momentum within NASA. At a meeting of our group in April 1960—it was now called the Lunar Science Subcommittee of the Space Steering Committee, and I was still its chairman—Homer Newell made a striking statement that the lunar exploration was the second highest program in NASA next to Mercury.

Up to this point there had not been very much discussion of manned versus unmanned landings; our planning was centered on the unmanned missions. In a meeting of the lunar committee in May 1960, manned landings came into the discussion with a bang. Plans for Apollo included a circumlunar flight as a step toward the manned landings. The people in our group were very dubious about that entire proposition. They complained that the circumlunar flight had little or no scientific value and would take away scarce Saturn vehicles from the scientifically valuable soft landings. There was also a lot of complaining about the Mercury program. I never had any trouble in accepting Mercury. It was clear to me that our only chance of getting the science missions flown was to ride on the tails of the manned flight program. The taxpayers were not going to spend billions of dollars on the unmanned exploration of the moon just to satisfy the curiosity of a few people like Harold Urey and me about the origin of the solar system, and I did not see why they should.

However, the arguments continued, and, of course, some of the scientists became even angrier after Apollo got going.

Throughout this period the momentum continued to build in the lunar science program, but more and more as an adjunct to the manned flight program. The Lunar Science Committee was getting to be big business, and I could not handle the job and continue to build up the research in our Theoretical Division at the same time. Also, committee work was not my forte. I pulled out some time around the end of 1960.

LUNAR GEOLOGY

EUGENE SHOEMAKER

The systematic geologic mapping of the moon was born of a daydream in 1948. I had just gone to work for the US Geological Survey in the uranium country of western Colorado. The stockpile of Colorado Plateau uranium from the early radium mining days, which had been stored at Canonsburg, Pennsylvania, and also the uranium imported from the Belgian Congo had been expended in the construction of only three bombs. Now the AEC was gambling that enough new uranium ore could be found, not only to build more weapons, but also to start a nuclear power industry. Down at White Sands, New Mexico, von Braun's rocket group, which had escaped Peenemunde on a slow train just ahead of the advancing Russian army, was launching the remaining

Eugene Shoemaker is a geologist at the US Geological Survey, and professor of geology and planetary science at the California Institute of Technology. He has worked in the organization and administration of the US Geological Survey Branch of Astrogeology.

V-2's—built from the parts that had been packed on the train. To an impressionable 20-year old geologist, the late 1940s were very exciting times. Anything seemed possible. It even seemed possible that men would land on the moon during my professional career. Somehow, in my bones, I knew it was going to happen and made up my mind to be standing at the head of the line to be one of those men. Why else would anyone go to the moon, except to explore the geology? The ultimate sequence of events that would bring the nation into the Apollo program was beyond my imagination, but not the means of getting in line.

Of course, if you told someone in 1948 that you were planning to do geologic field work on the moon, you would have been considered a prime candidate for the lunatic asylum. The goal remained a private one for several years. It seemed likely to me then that the National Academy of Sciences would, at the appropriate time, be called upon to form an *ad hoc* committee to review the qualifications of scientists to be sent to the moon. Sixteen years later, this premoni-

51

tion was fulfilled; by a curious twist of fate, no longer a candidate, I chaired that Academy *ad hoc* committee.

In order to be at the head of the line, my plan in 1948 was to become the most experienced field geologist working on problems related to the surface of the moon. The first turn onto the road to the moon came in 1952 when, quite by accident, I discovered low grade uranium deposits in the Hopi Buttes of Arizona.[13] This afforded an opportunity to return to the Hopi Buttes, where more dissected maar-type volcanoes are exposed than in any other place known in the world. The original surface form of this type of volcano resembles more closely the small craters of the moon than do other volcanic craters. The chance to study these closely and to decipher their origin was an unusual stroke of luck.

While I was working in the Hopi Buttes, another unusual opportunity presented itself. At Los Alamos, an extraordinarily gifted young designer of nuclear weapons had a novel idea to improve the plutonium supply. Why not create a plutonium deposit, by wrapping a nuclear device in a uranium blanket and exploding the device underground? A lot of plutonium could be made in a hurry, and it might not be too difficult to mine it out, particularly if the whole experiment could be done in ice. Attention soon shifted to salt as the preferred medium, but the project was code-named MICE. Its creator was Theodore Taylor who, a few years later, conceived Orion, a nuclear-propelled rocket ship capable of taking man to the outer planets.[11]

The AEC came to the Geological Survey for advice on large bodies of ice, salt, and other potential host rocks for a synthetic plutonium deposit. Almost as an afterthought, I became involved in MICE because I was working on "explosive" volcanoes; it seemed worth checking whether the plutonium or plutonium laden rock might be "erupted" to the surface. After looking into it, containment did not seem to

me to be a serious difficulty; dispersal of the plutonium in a large body of broken rock looked like a possible consequence that could pose practical problems of extraction, however. I decided to map the craters produced by two shallow subsurface nuclear explosions at Yucca Flat, Nevada, in order to see how the fission products were dispersed.[14] I also went to Meteor Crater, Arizona, to see if the phenomena of the kiloton craters at Yucca Flat could be scaled up to the metagon range. Somewhat to my surprise, Meteor Crater turned out to be an eerily close scaled up version of the Teapot Ess nuclear crater. A detailed study of Meteor Crater seemed appropriate.

The moon, of course, was never far away from my thoughts in doing this work. About the time I started working on MICE in 1956, I approached the Director of the Geological Survey, Thomas B. Nolan, to see whether it might be possible to initiate a study of the geology of the moon after completing the Hopi Buttes work. I had in mind, roughly, a four-man project to start mapping the geology of the moon with the aid of the best quality telescopic photographs, supplemented with direct visual telescopic observations. Nolan took the proposal quite seriously and encouraged me to pursue planning the work. He steered me to William W. Rubey, then at the Survey, who was also encouraging, and helped by checking whether a similar effort might already be underway on a classified basis, in the Defense Department. There was not. It is a credit to the foresight of Nolan and Rubey that they thought the subject worthwhile. This was a full year before Sputnik. Perhaps the legacy of G.K. Gilbert's great pioneering study of the moon, done while he was Chief Geologist of the Survey,[3] had paved the way for their acceptance of this strange proposal from a brash young man.

Sputnik was launched while I was enroute from a MICE project meeting at Oak Ridge, Tennessee, back to my field camp in the Hopi Buttes. We learned about it over the radio in camp. It seemed plain, then, that the race to the moon would surely be on. I was midway in detailed mapping at Meteor Crater[15,18] and midway in the studies of the Hopi Buttes.[26] Neither project would be finished to my satisfaction. But they did provide a solid foundation for evaluating the origin of lunar craters.[17]

The next year, the uranium exploration program in the Geological Survey was terminated, and I moved to the Menlo Park, California, office of the Survey. With the aid of some exceptional photographs of the moon taken with the 100-inch Hooker telescope at Mount Wilson Observatory, I started serious work on the lunar craters, while writing up the Arizona and Yucca Flat field studies. These photographs included the highest resolution pictures taken of the moon up to that time and could be purchased from a catalogue at the Caltech bookstore. At about the same time, G.P. Kuiper and his colleagues had assembled the best photographs covering the entire earth-facing side of the moon and prepared them for publication by the U.S. Air Force.[7]

At the instigation of John O'Keefe, the U.S. Army Corps of Engineers had also become interested in the moon, and funded a study by the Military Geology Branch of the Geological Survey in 1959. This study was carried out by Arnold C. Mason and Robert J. Hackman, in Washington, D.C., who prepared a preliminary small-scale geologic and terrain map of the subearth face of the moon.[6,10] On the west coast, I did some scouting for information for the Military Geology Branch and became a regular attendee of the somewhat informal Lunar and Planetary Exploration Colloquium in the Los Angeles area.

Through contacts with Albert R. Hibbs and Manfred Eimer of Caltech's Jet Propulsion Laboratory, I learned of a proposed new effort by the Air Force Chart and Information Center to start topographic mapping of the moon at a scale of 1:1,000,000. This effort was an outgrowth of the Kuiper

photographic atlas. Their first prototype map included the crater Copernicus, and I used a copy made available by Eimer as a base for compilation of a first prototype lunar geologic quadrangle map. A trial printing was then made of the combined geology and topography through the courtesy of Robert W. Carder of ACIC, in St. Louis, Missouri. Although still quite preliminary in a number of respects, this first quadrangle map showed that the geology could be solved and represented cartographically. It formed the basis for a proposal to NASA to initiate a detailed geologic mapping program, which was warmly and effectively supported by Eimer and O'Keefe.

In the Spring of 1960 a remarkable new mineral called coesite, a high pressure polymorph of silica, was discovered in samples I had collected from Meteor Crater.[1] This new mineral provided us with a diagnostic tool for the recognition of other impact craters and structures. We applied it immediately to demonstrate the impact origin of the great Ries crater in Germany.[21] It is safe to say that this constituted a breakthrough in the recognition and understanding of the impact history of the earth. The discovery of coesite was extremely timely; it helped persuade NASA of the merits of our proposed geologic mapping of the moon. The program was funded and officially started in August 1960.

Before this official start, however, Hackman and I had joined forces to refine the Copernicus map. We·showed that it was possible to decipher the stratigraphy of the moon, using the relations revealed in the very best telescopic photographs. Age and superposition relationships could be solved by paying close attention to the distribution of secondary craters. An overall stratigraphic framework was obtained for the center of the moon's face that served as the basis for a successful lunar geologic timescale.[22] The details of the secondary crater pattern of Copernicus and their implications for the impact origin of the primary crater were first published in the proceedings of the Lunar and Planetary Exploration Colloquium.[16] While geologic terrain maps had been published previously,[27] this was the first map that dealt with the detailed stratigraphy in the way that led to unambiguous conclusions about the sequence of events and geologic history of a region of the moon.

In the Fall of 1960, Richard E. Eggleton and Charles H. Marshall joined the new project. We set ourselves to map a block of four quadrangles that covered the target area for landing the Ranger spacecraft. This was the area assigned first priority by the Jet Propulsion Laboratory for topographic mapping by ACIC. The Kepler quadrangle was done by Hackman,[4] the Letronne quadrangle by Marshall,[9] and the Riphaeus Mountains quadrangle by Eggleton.[2] The Copernicus quadrangle was ultimately redone and refined by Harrison H. Schmitt, Newell J. Trask, and Shoemaker.[12] From analysis of the crater density on the major lunar stratigraphic units and from the age and distribution of ancient impact structures in the United States, we derived a first approximate absolute chronology of the lunar geologic timescale.[23] This first rough chronology has withstood the acid test of isotopic dating of returned lunar samples surprisingly well.

Mapping of the four quadrangles introduced many new refinements in our understanding of the stratigraphy. Eggleton first recognized the widespread plains unit now called the Cayley Formation. At an early stage, he and I both recognized it might be a distant deposit of mobile ejecta from one or more of the late impact basins. Somehow, this early interpretation was forgotten by the geologists who later studied the planned landing site for Apollo 16, but it proved to be the right answer. He also mapped and named the Fra Mauro Formation and showed that it corresponded to a great blanket, of tapering thickness, around the Imbrium Basin. Marshall[8] solved for the thickness of mare lavas in the Oceanus Procel-

larum. In a later map, Hackman[5] recognized and separated the craters of the Archimedian Series, now included in the Imbrian System.

The Branch of Astrogeology at the survey was organized in 1961, and the lunar geologic mapping team was soon joined by Harold Masursky, Henry T. Moore, Elliot C. Morris, John F. McCauley, Michael H. Carr, and Don E. Wilhelms. Ultimately, several dozen geologists of the expanded Branch contributed to mapping 44 quadrangles covering the subearth side of the moon. Early on, we knew that direct observations at the telescope were essential to solve the stratigraphic relationships. Key to this work was the study of secondary craters, which were commonly at or below the threshold of resolution in the best available photographs. Under very favorable conditions of seeing, we could observe at the telescope secondary craters four times smaller than those resolved in the photographs. The work was done, initially, at the 36-inch refractor of the Lick Observatory at Mount Hamilton, California, and at the 24-inch refractor at the Lowell Observatory in Flagstaff, Arizona. In 1963, we designed and placed into operation a new 30-inch Cassegrain telescope at Flagstaff that was dedicated to the lunar geologic mapping program. This also precipitated a move of the headquarters of the Astrogeology Branch to Flagstaff.

In 1964, a long series of attempts to secure close-up television images of the moon with the Ranger spacecraft finally succeeded. These images revealed the population of small craters on the moon, down to about one meter diameter, and provided the initial observational foundation for analysis of a thin veneer of fragmental debris of impact origin, the lunar regolith, which cloaks the lunar surface.[19,20] The first of five successful landings of the Surveyor spacecraft took place in 1966. These landings gave an intimate picture of the structure and physical properties of the regolith[24,25] and the first direct measurement of its bulk chemical composition. From the landing sites on the lunar maria the basaltic composition of the mare lavas, which we had deduced from the form of the lava domes, was confirmed. The first of five successful Lunar Orbiter spacecraft was also flown in 1966. From the Lunar Orbiter IV, flown in 1967, high resolution photographic coverage was obtained for almost the entire near side. This coverage, with ten times better resolution than the telescopic photographs, revolutionized the geologic mapping progran. Both the accuracy of the stratigraphy and the speed with which the work could be done were greatly enhanced. Lunar Orbiter V took still higher-resolution photographs of targets of special interest selected by our geologic mapping team. From these selected areas, the landing sites of the later Apollo missions were eventually chosen. Some of the key individuals in the analysis of the Lunar Orbiter photographs were Masursky, McCauley, and Wilhelms. A preliminary compilation of the geology of the subearth side of the moon was published by Wilhelms and McCauley.[29] A final monographic synthesis of the entire geologic mapping effort is being completed by Wilhelms.[28]

The exploration of the moon culminated in the manned landings of Apollo 11 through Apollo 17. Just one man from our lunar geologic mapping program arrived to carry out field work on the moon. He was Harrison H. Schmitt. Here, I will confess to mixed feelings as I watched Schmitt leave earth in the spectacular night launch of Apollo 17 at Cape Canaveral. There were feelings of deep satisfaction and pride that Schmitt was in that spacecraft. On the other hand, had it not been for a defunct adrenal gland—gone bad in 1962—it might have been me. Youthful dreams die hard. They died hard for some of the others on our astrogeology team too, especially Daniel J. Milton and Michael B. Duke, who passed the Academy's scrutiny but not NASA's incredibly stiff physical exam.

The return of lunar samples by Project Apollo resulted in an unprecedented scientific bonanza. New techniques were applied to the analysis of the lunar rocks and better analyses were done than had ever been obtained for the rocks on earth. From a standpoint of global geology, perhaps the most exciting returns were the high-precision isotopic age determinations. Crystallization ages and shock metamorphic ages were obtained that gave us many precise points on the lunar geologic time scale. Although our original estimates were roughly right, the true ages differed in very important respects. Most notably, it was found that the visible great impact basins of the moon were not formed right at the beginning of lunar history, but some 400 to 700 million years later. It has taken nearly a decade to unravel the full meaning of this result. The timing of the late heavy bombardment is due to the long dynamic lifetimes of several classes of bodies—planetesimals—from which we think the planets were formed. It now appears that most of the last planetesimals of the earth and planetesimals of Uranus and Neptune remaining on planet-crossing orbits were finally swept up at about 4.0 billion years ago. Eggleton's Fra Mauro Formation and Wilhelms' Cayley Formation are regional deposits on the moon associated with the last two big planetesimals to strike the moon.

REFERENCES

1. Chao, E.C.T., Shoemaker, E.M., and Madsen, B.M., "First natural occurrence of coesite from Meteor Crater, Arizona," *Science, 132,* 220-222, 1960.

2. Eggelton, R.E., Geologic map of the Riphaeus Mountains region of the Moon: U.S. Geol. Survey Map I-458, 1965.

3. Gilbert, G.K., "The Moon's face; a study of the origin of its features," *Bull. Phil. Soc. Wash. 12,* 241-292, 1893.

4. Hackman, R.J., Geologic Map and sections of the Kepler region of the Moon: U.S. Geol. Survey Map I-355, 1962.

5. Hackman, R.J., Geologic map of the Montes Apenninus region of the Moon: U.S. Geol. Survey Map I-463, 1966.

6. Hackman, R.J., and Mason, A.C., Engineer special study of the surface of the Moon, U.S. Geological Survey, Misc. Geologic Investigations, Map I-351, 1961.

7. Kuiper, G.P., Lunar Atlas, U.S. Air Force, Vol. 1, 1960.

8. Marshall, C.H., "Thickness of the Procellarian System, Letronne region of the Moon," Geol. Soc. America Spec. Paper 68, 225, 1962.

9. Marshall, C.H., Geologic map and sections of the Letronne Region of the Moon, U.S. Geological Survey, Geology of the Moon Letronne Region, I-385 (LAC-75), 1963.

10. Mason, A.C., and Hackman, R.J., "Photogeologic Study of the Moon," *The Moon,* eds. Z. Kopal and Z.K. Mikhailov (London: Academic Press, 1962), pp. 301-315.

11. McPhee, John, *The Curve of Binding Energy* (NY: Ballantine, 1973).

12. Schmitt, H.H., Trask, N.J., and Shoemaker, E.M., Geologic map of the Copernicus quadrangle of the Moon, U.S. Geological Survey Map I-515 (LAC-58), 1967.

13. Shoemaker, E.M., "Occurrence of uranium in diatremes on the Navajo and Hopi Reservations, Arizona, New Mexico, and Utah," in United Nations, Geology of Uranium and Thorium: Internat. Conf. Peaceful Uses Atomic Energy, Geneva, Aug. 1955, Proc., *6,* pp. 412-417,1956. Slightly revised, in Page, L.R., "Contributions to the geology of uranium and thorium . . . ", U.S. Geol. Survey Prof. Paper 300, pp. 179-185.

14. Shoemaker, E.M., "Brecciation and mixing of rock by strong shock," Art. 192 U.S. Geol. Survey Prof. Paper 400-B, pp. B423-B425, 1960.

15. Shoemaker, E.M., "Penetration mechanics of high velocity meteorites, illustrated by Meteor Crater, Arizona," Internat. Geol. Cong., 21st, Copenhagen 1960, Report, pt. 18, pp. 418-434, 1960.

16. Shoemaker, E.M., "Ballistics of the Copernican ray system," Lunar and Planetary Exploration Colloquium Proc., *2,* no. 2, pp. 7-21, 1960.

17. Shoemaker, E.M., "Interpretation of lunar craters," in Kopal, Zdenek, ed., *Physics and Astronomy of the Moon* (London: Academic Press, 1962), pp. 283-359.

18. Shoemaker, E.M., "Impact mechanics at Meteor Crater, Arizona," in Middlehurst, Barbara, and Kuiper, eds., *The Moon, Meteorites, and Comets—The Solar System,* Vol. IV (Chicago: Univ. Chicago Press, 1963), pp. 301-336.

19. Shoemaker, E.M., "Interpretation of the small craters of the Moon's surface revealed by Ranger VII," in Transactions of the International Astronomical Union, Proceedings of the 12th General Assembly, Hamburg, 1964, vol. XIIB, pp. 662-672, 1966.

20. Shoemaker, E.M., "Preliminary analysis of the fine structure of the lunar surface in Mare Cognitum," in Hess, W.N., Menzel, D.H., and O'Keefe, J.A., eds., *The Nature of the Lunar Surface; Proceedings of the 1965 IAU-NASA Symposium* (Baltimore: Johns Hopkins, 1966), pp. 23-78.

21. Shoemaker, E.M., and Chao, E.C.T., "New evidence for the impact origin of the Ries Basin, Bavaria, Germany," *Jour. Geophy. Research, 66,* 3371-3378, 1961.

22. Shoemaker, E.M., and Hackman, R.J., "Stratigraphic basis for a lunar time scale," in Kopal, Zdenek and Mikhailov, Z.K., eds., *The Moon,* pp. 289-300.

23. Shoemaker, E.M., Hackman, R.J., and Eggleton, R.E., "Interplanetary correlation of geologic time," *Advances in the Astronomical Sciences, 8* (New York: Plenum Press, 1963), pp. 70-89.

24. Shoemaker, E.M., and Morris, E.C., "Physical characteristics of the lunar regolith determined from Surveyor television observations," *Radio Science, 5,* 129-155.

25. Shoemaker, E.M., and Morris, E.C., "Surveyor Final Reports—Geology: Craters: Fragmental Debris; Fragmental Debris Physics," *Icarus, 12,* 167-212, 1970.

26. Shoemaker, E.M., Roach, C.H., and Byers, F.M. Jr., "Diatrenes and uranium deposits in the Hopi Buttes, Arizona," *Petrologic Studies* (Geol. Soc. America, 1962), pp. 327-355.

27. Spurr, J.E., *Geology Applied to Selenolog: Vol 1,The Imrian Plain Region of the Moon* (Lancaster Pa.: The Science Press, 1944).

28. Wilhelms, D.E., "Geologic History of the Moon," U.S. Geol. Survey Prof. Paper (in preparation).

29. Wilhelms, D.E., and McCauley, J.F., Geologic map of the near side of the Moon: U.S. Geol. Survey Map I-703, 1971.

OBSERVATIONS OF HIGH INTENSITY RADIATION BY SATELLITES 1958 ALPHA AND 1958 GAMMA

JAMES A. VAN ALLEN

The successful launch on January 31, 1958, of the first American satellite, Explorer I, placed into orbit a radiation detector which at first seemed to malfunction. In certain regions of the orbit the instrument ceased to count charged particles. A careful analysis by James Van Allen and his colleagues revealed not a failure but the discovery of belts of radiation trapped in the earth's magnetic field that were so intense as to "saturate" the instrument—radiation far stronger than anyone had imagined. The following paper is an edited transcription of Dr. Van Allen's first presentation of his unexpected results to a major national audience, a joint symposium of the American Physical Society and the National Academy of Sciences, on May 1, 1958. We reprint it here as a record of the detective work that infused this exciting discovery of 1958. It appeared in slightly different form as "Satellites 1958 Alpha and Gamma, High Intensity Radiation Research and Instrumentation," IGY Satellite Report Number 13, January 1961, National Academy of Sciences - National Research Council, Washington, D.C.

I am making a report on behalf of my colleagues Mr. Ludwig, Mr. McIlwain, Dr. Ray and myself. In spite of the title of the program the major portion of the results I am reporting this morning comes from 1958 Gamma or Explorer III, although we have also results which I will describe from 1958 Alpha as well. I should mention at the outset that obtaining results in this satellite field is quite a large operational undertaking, and I should particularly like to mention the fact that these results are, of course, quite impossible without the great IGY meshwork of activities and agencies, in particular the Jet Propulsion Laboratory in Pasadena, the Army Ballistic Missile Agency in Huntsville, Alabama, and the Vanguard Group at the

James A. Van Allen is now Carver Professor of Physics and Head of the Department of Physics and Astronomy at the University of Iowa, Iowa City.

58

Naval Research Laboratory.

The object of this experiment as it was planned about two years ago was to make a comprehensive and precise survey of cosmic ray intensity as a function of latitude, longitude, altitude and time; in short, to measure omnidirectional intensity of cosmic rays as a function of r, φ, θ and t, and then to subject this three-dimensional network of intensities to a detailed analysis for the symmetry in the earth's magnetic field, for the altitude variation, and for the primary spectrum, and for the discussion of the intensity and latitude distribution of albedo.

At the present time we have had data successfully from 1958 Alpha for about a six-weeks period. We have quite a large quantity of results. In Alpha we had a continuous telemetering system from which the data could be obtained only during the passage nearby one of the sixteen receiving stations. On each pass we obtained typically a minute or two minutes of workable data. So the results, although quite worthwhile and present now in quite large quantity, are none-the-less relatively fragmentary in terms of the total potential results, total intensity to which the satellite is subjected during its lifetime. In Gamma we have the full package which we originally intended. It contains a magnetic tape storage device which collects data throughout the entire orbit around the earth. At the present moment we have about five weeks of observations, the apparatus is still going strong, and it has an anticipated death date of around the first of June. So, during this period of time we expect to obtain about 720 round trips around the earth. Based on the present Minitrack batting average, we will likely have about five to six hundred sets of usable data, each one giving the intensity at each point in space in a circuit around the earth.

I give at this time a brief summary of results and then I'll supply the substantiating detail. The cosmic ray survey aspect of the work is performing quite well; we are getting the results anticipated in a general way from the analysis of the expectation from this experiment. The detailed analysis of the cosmic ray intensity is a rather lengthy job and I don't think I will report that at this time, although we have this already under way. The main point of the paper this morning is the following:

Above an altitude of about one thousand kilometers we have encountered a very great increase in radiation intensity which is vastly beyond what could be due to cosmic rays alone. In fact, it is of the order of, or greater than, one thousand times the intensity of cosmic rays as extrapolated to these greater altitudes, and it's this topic that's the main subject of my discussion this morning.

As you see from Figure 1, the apparatus on board these birds is quite complex and this is only a relatively simplified block diagram of the apparatus on board Explorer III, or 1958 Gamma. I also point out the similarity to Alpha. So far as we are concerned at Iowa, the main elements are the following: a single Geiger counter, a factor of 32 scaling circuit, a subcarrier oscillator and a continuously operating low-power transmitter, the Microlock transmitter of the JPL. In addition, this is scaled by an additional factor of 4 so the Geiger counter rate is scaled down by a factor of 128. It goes to what we call an inhibitor or anticoincidence circuit. In a parallel channel of electronics we have a tuning fork which is the on-board time standard that enables us to determine from the result of an interrogation record the point in space at which the data were being obtained. This tuning fork is scaled by a factor of 512 so that it produces 1 pulse per second, and then by virtue of the inhibitor circuit, whose operation I will show in a moment, we are able to knock pulses out of the series of time, the chain of time-pulses with the 128 Geiger counts. This is recorded on a magnetic amplifier, a magnetic recorder, tape recorder, and then eventually stored up. The storage time is 135 minutes or a little over an orbital period. Then there is the Vanguard command receiver on board. Al-

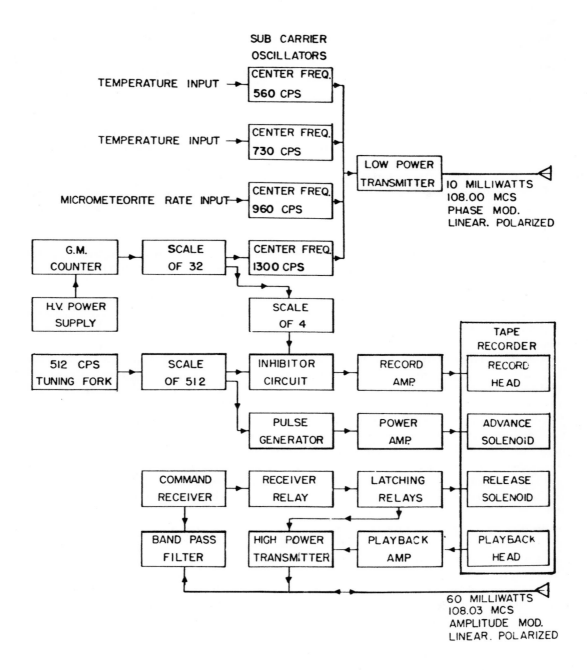

Figure 1

though this is usually called a command receiver, I decided lately I should call it a "request receiver" because we are not in a position to take any disciplinary action if it doesn't respond. This receiver then unlatches the magnetic tape which plays back and at the same time turns on the higher powered Minitrack transmitter which radiates for about a six-second period and then is turned off. So on the occasion of a successful request we receive the entire orbital data in about a six-second period played back on a magnetic recorder at one of the Minitrack ground stations.

Figure 2 illustrates the action of the circuit. These (A) represent one-second time pulses in original recording time. This (B) would be, for example, a set of scaled Geiger pulses and the inhibitor circuit acts so that one cancels off the corresponding time pulse, so the final recorded record on the magnetic tape looks like this(C): pulse, pulse, missing tooth, series of pulses, another missing tooth, and so forth.

Figure 3 shows the apparatus in Explorer III. The main element of interest at the moment is our Geiger counter: an Anton, halogen-quenched counter. This (A) is chuck full of electronic components and contains among other things the command receiver (B), the high-powered transmitter,

the tape recorder, all of the batteries and associated electronic circuits. This is the low-power transmitter here. These whiskers (C), were taken off in the final flight version in order to reduce the tendency toward going into a tumble. In spite of this, tumbling did ensue after about a week.

Figure 4 shows a sample clipped out of one of the played-back records. This particular one came from a San Diego interrogation. You see how life is going here. The data shown here were obtained at a position about 1000 miles southwest of San Diego, out over the Pacific Ocean and at a position ordinarily quite inaccessible to observation. Here (beginning at A) you'll see there are one, two, three, four, five tuning fork pulses present, then a missing one; then one, two, three, four, five, six and a missing, and so forth. So, at this particular time, the counting rate of the counter was about one-sixth of 128, or about 20 counts, per second. This illustrates the mode of reduction of the data.

Figure 5 is an old slide just to remind you how cosmic ray intensity goes in rocket experiments. It increases from the ground upwards to the transition maximum, then comes down again at around 50 kilometers and up to about 120 kilometers. In this particular flight the intensity was substantially

Figure 2

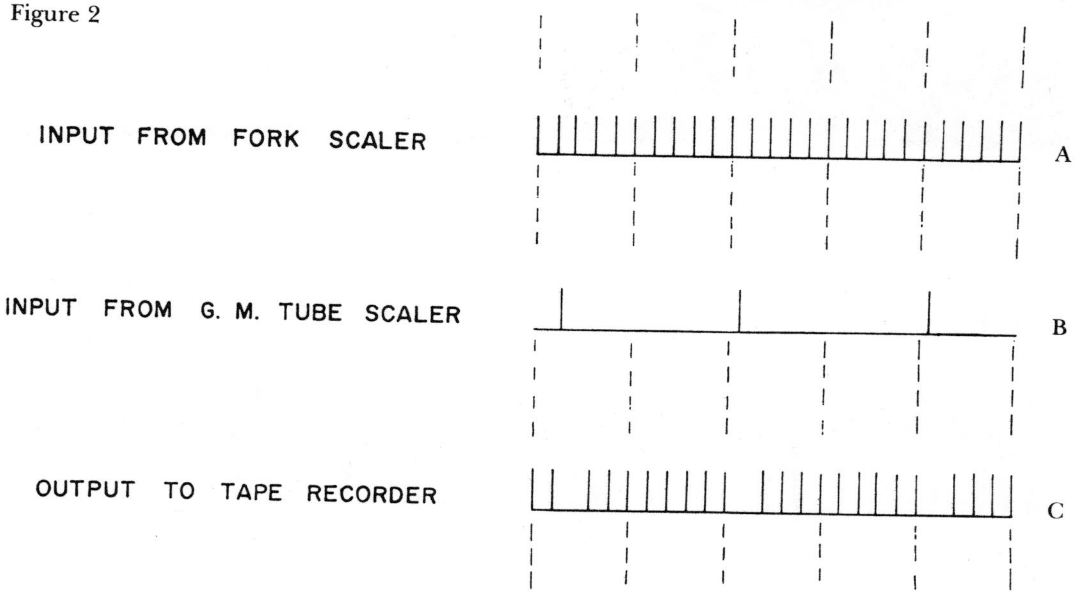

INPUT FROM FORK SCALER — A

INPUT FROM G. M. TUBE SCALER — B

OUTPUT TO TAPE RECORDER — C

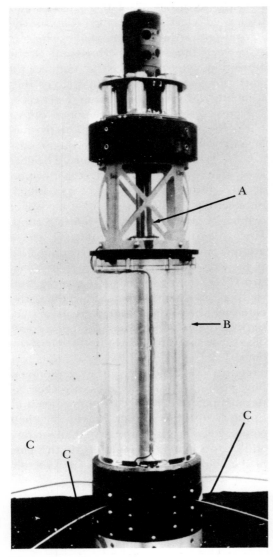

Figure 3

Figure 4

constant. However, as we have known for a long time, one cannot expect this plateau to be of indefinite altitude duration, and in the next slide I show how we have anticipated it would change. Figure 6 is also a rather old slide, vintage about 1949, in which we made some estimates of how we might expect the intensity to depend upon altitude if there ever was a satellite. The parameter in this curve is geomagnetic latitude. These results are crude theoretical calculations, in other words, sort of an estimate of what might happen. Let me point out, for example, at 45 degrees geomagnetic latitude, (the horizontal scale, is one taking the radius of the earth as unity) the whole curve would be crammed into this little gap (A). At 45 degrees we expect the intensity to increase rather rapidly as a function of altitude. This (B) would be at one-and-a-half earth radii. Explorers I and III have had an apogee of about 1.4 earth radii, so this (C) would be approximately that point.

Figure 7 is an experimental curve taken from the data of Alpha, and all of these points come from the Microlock stations* over southern California. The horizontal axis is the height in kilometers, zero to 1600 kilometers. The small circles are observed intensities—counting rate in counts per second. The straight line is the experimentally-observed dependence of counting rate on altitude, and it is in quite sensible agreement with the expectation. The scatter of points is primarily due, I believe, to the fact that the geomagnetic latitude is not the same for all of these points even though they are all obtained from data within a radius of about 50

*A Microlock station is one of several special phase-lock-loop telemetry receiving stations established by JPL for the early Explorer missions. They were in addition to the worldwide distribution of NRL Minitrack stations.

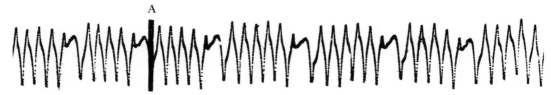

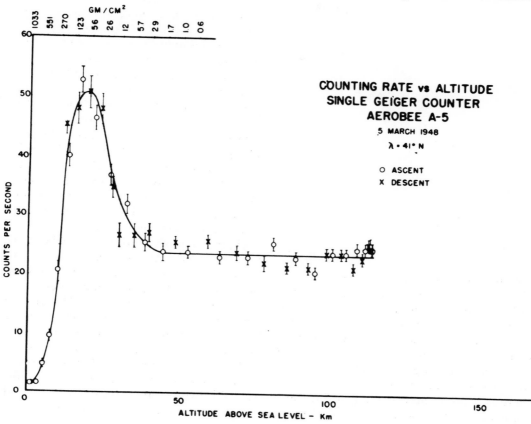

COUNTING RATE vs ALTITUDE
SINGLE GEIGER COUNTER
AEROBEE A-5

5 MARCH 1948

λ • 41° N

O ASCENT
X DESCENT

Figure 5

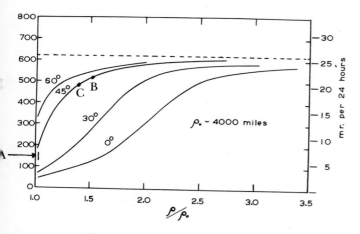

Figure 6

miles. The various passes by these stations varied considerably in impact parameter so the geomagnetic latitudes are not the same, and we have not yet taken that effect out of the point, so to speak, so these are raw points. All the points were obtained over southern California during the first two weeks of 58 Alpha.

Figure 8 shows another feature of the results of 58 Alpha, which at first we regarded with a very jaundiced eye. However, I think we are forced to believe in them, and we have a very great body of results from Alpha which support these observations, and the thing that has made it, I think, a completely certain result is the vast amount of data we are getting from Gamma in which we get complete orbital playback. This is a latitude-longitude plot and there for example is Quito, the location of one of the Minitrack

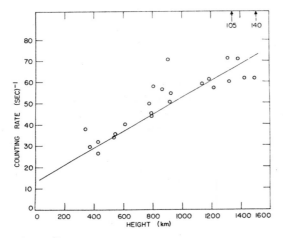

Figure 7

stations from which we have gotten a good bit of data. This figure was made up in the rather early days of the reduction of the Alpha data. The arrows indicate the direction of the orbit; they are the direction of motion of the satellite in its orbit as projected on the latitude-longitude plot. All of these heavy lines show that, so far as we could tell, there were no counts occurring during these segments of the orbit, and I'll elaborate on that in a moment. The most immediate interpretation is that there is no radiation intensity there. However, we are quite sure that's wrong and it is really that it is due to the fact that the counter was blocked due to very high radiation intensity. On this slide we have two interesting transition cases (A, B). We have a considerable number more now since this slide was made, of this sort (A), where there were no counts and then as it ran along it started counting again, then continued counting rather rapidly. And here (B) is another very interesting case near Antofagasta, one of the other Minitrack stations, in which the altitude was changing. All of these were at about 2200 kilometers altitude—these five cases. The case near Antofagasta was at about 1100 kilometers at the start and about 1400 at the end of the recorded pass; the counting rate was more or less sensible-looking at the start and then

blanked as it came along and this transition occurred at about 1100 kilometers.

Figure 9 is another way of looking at the same type of data. Here we have height as a vertical scale and geographic latitude horizontally. Each one of these black points represents a sample of the radiation intensity in which apparently no counts occurred during the two-minute period during the pass. This point (A) is one with a transition point which I just showed. The four points at a lower altitude showed counts at quite sensible cosmic ray intensity levels of the order of 30 counts per second. The point (A) is at about 1200 kilometers altitude.

Figure 10 shows the results of a laboratory test on the 1958 Alpha apparatus and it's to illustrate the interpretation which we have now given to the observations. The curve represents the counting rate of our real counter as a function of the counting rate of a counter which has zero dead time. This you see is a semi-log plot which gives it a rather peculiar appearance but is necessary to show the results properly on one slide. Actually if you read off the curve you'll find that the counting rate of our real counter

Figure 8

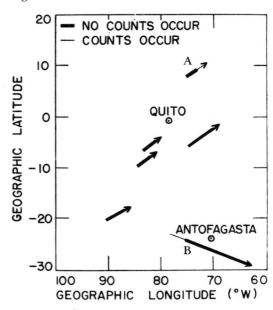

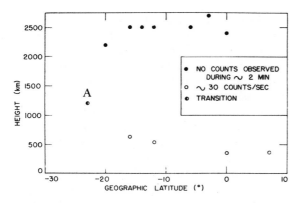

Figure 9

does follow the counting rate of an idealized zero dead-time counter quite well up to around 300 counts per second. This graph for the present purposes goes off the top of the slide and the significance of that is we are unable, in either Alpha or Gamma, to follow in detail counting rates exceeding about a rate something like in here (above A). In Gamma we cannot follow in detail a counting rate greater than 128 per second, as you can easily judge from the previous discussion of the system; in Alpha we can go somewhat higher on the continuous channel into about 200 to 250 per second. As one increases the radiation intensity, one finds that the counting rate observed comes back down. The reason for this is that the dead time of the counter is such that for a sufficiently high rate no output pulse from the Geiger counter is of sufficient amplitude to trigger the scaling circuit, so that eventually at about 2500 counts per second no counts appear from the system from either Alpha or Gamma, which is as I've just described. In the meanwhile there is a transition region when there are some pulses large enough, but only a relatively small fraction of the total number which would be reported by a zero dead-time counter. This means that an observed counting rate of, for example, 50 counts per second, may correspond to this radiation intensity (B) or it may correspond to this radiation intensity (C). So the observed output is a double-valued function of

the true radiation intensity.

Now Figure 11 shows the first playback which we obtained for study from 58 Gamma. It was one obtained from an interrogation over San Diego. The actual raw results are shown here. This is counted from the time of the previous interrogation, which was a Fort Stewart interrogation in Georgia. The counting rate was about 20 counts per second; it diminished somewhat to about 16 counts per second, then made a very rapid increase and went off the map here (B).

In here (around A) we had on the order of one missing tuning-fork pulse out of six, and in here (above A) we started having more and more missing pulses. At this point (B) they were all missing and the only being played back was the noise level of the receiver. From here (B) to here (C) we had no pulses appearing. At this time (C) there was a rather rapid transition down to a circumstance in which every tuning fork pulse was appearing and this corresponds to a rate exceeding about 25,000 per second. Then at a later time (just before D) it came back on the scale, so to speak, and then all the pulses were missing again (at D) for this length of time (about the next 50 minutes) and then it came back down again to a quite reasonable cosmic ray value here (E). As you know, the perigee of Gamma was rather near the most northerly point during the first few days of the flight which was launched on the 26th of March. This was about two days later. So the

Figure 10

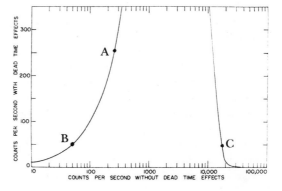

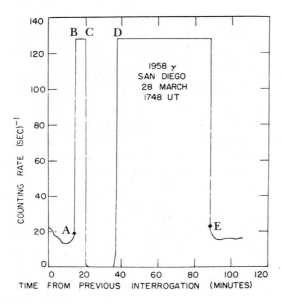

Figure 11

est altitude. The altitude increased mono-tonically as a function of latitude going southward and had its greatest value, its apogee value, near the lowest point. I might remark also when we made this slide we had a considerable uncertainty about the posi-tion of the satellite in space because we had at that time no reduced orbital data; we as-sumed the perigee at the most northerly point, and we made a rudimentary calcula-tion on the orbit for the purpose of plotting these points. Nonetheless, I think, it is quite clear even from this preliminary reduction that in a general way the intensity is more or less sensibly cosmic ray in value near the low altitude upper part of this graph; near the lower part we have these black segments which represent a very high counting rate, and these are also the highest altitude data. The transition occurs some place in through here (A for example) and there is a sugges-tion of a very interesting longitude effect as follows: that there seems to be a greater tendency toward higher counting rates in an easterly longitude of something like 300 to 340 degrees. This is roughly the antipode of the eccentric dipole of the earth, which is located at about 160 degrees east longitude, and it does appear that this may be a sugges-tive result of considerable interest, namely, that at a given geometric altitude one is at a lower geomagnetic altitude in this longitude.

data here (before B) correspond to the low altitude part of the orbit near perigee. This part of the curve (before B and D) should actually be, so to speak, turned inside out and would go up about 100 times as high as the top of this graph and stay up above this value for a short time and then come back down to a rate intermediate between 15,000 per second and 128 per second.

I will show you a summary of nine differ-ent passes around the earth that was made about two weeks ago. (Figure 12) We now have considerably more data. This is a latitude-longitude plot of the subsatellite point. Along each individual orbit is shown in a draftsman's sort of code what was going on. In the regions of the plain lines we had a normal cosmic ray rate, less than 128. The segments having the little circles on them represent a rate between 128 and 15,000. The heavy black parts represent a rate ex-ceeding 15,000 and in most cases exceeding 25,000. I remind you that near the upper end of the orbit at the highest northerly latitude, about 33 degrees, during this period of 28 March to 31 March the satellite was very near perigee, and that was the low-

This is more in the nature of an inter-pretative speculation from this point on. I think I'll write on the board in a moment the broad conclusions we have. As mentioned before, particularly by Dr. Siry,* this orbit has been a rather nasty one to reduce prop-erly because of the low perigee and the vari-able aspect and the rather high drag. So we do not yet have a very satisfactory orbit avail-able. However, we have a pretty respectable preliminary orbit. I plotted in addition to those cases shown just recently another ten cases, and they all show repetitively this same

---

*Dr. Siry was a member of the NRL Vanguard group in charge of the Minitrack telemetry stations and orbital calculations from observed data.

effect. In addition, we have examined in rough outline about 25 more cases and every one of them shows the repetitive performance as illustrated here, namely, a low counting rate of the order of 20 to 100 counts per second near perigee; a very high rate, exceeding 25,000 per second, at the highest altitude. And, of course, the orbit precesses and moves about in a very beautiful way for obtaining a three-dimensional meshwork of data. Now we have the fact that the floor for this effect is about 1000 kilometers and, as we've determined it so far, is roughly independent of latitude and does exist at all latitudes covered by the path of the satellite at the present time. I think that it is clear what we have. These are the raw data basically I have reported so far, and there is a very

rich field, I believe, for speculation about the significance and importance of these results. We believe that we are making *in situ* measurements on a real astrophysical plasma. The fact that the transition occurs rather sharply within an altitude range of about 100 kilometers and the fact that it does not extend below about 1000 kilometers appears to mean to us that there must be charged radiation involved, because the wall thickness of the counter is about 1.5 grams per square centimeter, that's including the shell of the satellite. Between 1000 kilometers and, let's say, 100 kilometers there is an amount of material in the atmosphere which is totally negligible in comparison to 1.5 grams per square centimeter. Hence, since the effect is confined to high altitude it cannot be due to

Figure 12

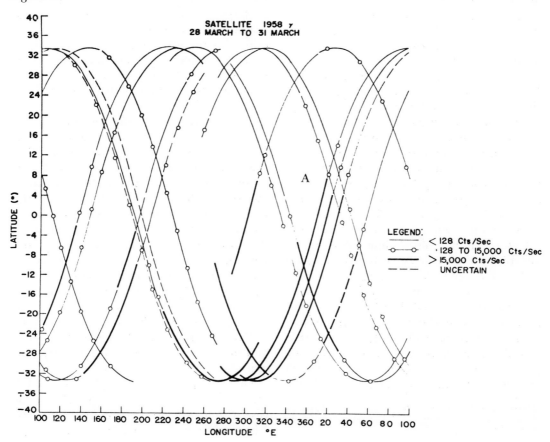

neutral radiations, so it must be due to charged particles. And the most plausible picture, we believe, is that what we are measuring is the bremsstrahlung from electrons which are a feature of this geophysical plasma and which are confined by the magnetic field from coming down any lower. Presumably, and I think I have said this is a speculative department at the moment, presumably these electrons rather closely resemble in energy and intensity—the intensity shows that it is reasonable to suppose that they are perhaps identical with and perhaps, at any rate, closely associated with—soft radiation which we have observed for the last several years beginning in 1953, in rocket measurements in the auroral zone and which have recently been observed in a way I'll show in a moment at balloon altitudes as well. So, as a rough picture of what's going on, we essentially adopt what has been given by more detailed measurements, particularly by the recent measurements of Mr. McIlwain at Iowa, on the nature, the energy spectrum and intensity of particles in aurorae.* He has successfully fired several Nike-Cajun rockets through aurorae with much more elaborate and discriminating detecting equipment. He finds, for example, that there are electrons present in aurorae with an energy spectrum extending out to about 150 kilovolts; the differential energy spectrum rises rather rapidly toward the zero and peaks at about 30 kilovolts and then appears to be dropping off at the instrumental limit of about 20 kilovolts. He also finds protons of a lower order of magnitude of intensity, but of a somewhat similar-appearing spectrum.

So based on this and the previous rockoon measurements in the Arctic and in the Antarctic we think that a plausible picture is that this plasma consists of electrons and likely protons, energies of the order of 100 kilovolts and down, mean energy probably

*McIlwain, C.E., "Direct Measurements of Particles Producing Visible Aurorae," *J. Geophys. Res., 65,* 2727-2747, 1960.

about 30 kilovolts, that these are circulating back and forth in the earth's magnetic field along lines of force. For example, the radius of curvature of a 100 kilovolt electron in the earth's field is the order of some tens or twenties of meters so that we think that the proper picture at the moment is that these particles are being reflected at the ends of lines as they go into stronger fields by the magnetic mirror effect so that they are circulating back and forth, and that they continue to circulate as long as their atmospheric penetration permits. We have undertaken to estimate the lifetime, and the lifetime at 1000 kilometers altitude is probably the order of hours. This is based naturally on very vague ideas of what the density is, since no one knows what the atmospheric density is, but a 100-kilovolt electron can wander around at 1000 kilometers for perhaps the order of an hour or several hours before running dead due to ionization. The intensity appears to increase quite rapidly with altitude. Of course how high it goes we don't know, and it may go up to vastly greater intensities than we have observed.

Let me show a few slides showing why we think this is a plausible connection. This is one of our old rocket flights from 1954 in the Arctic. (Figure 13) It shows how the intensity as measured with the Geiger counter increases with the ascent of the balloon, the rocket fires, then there is this very great increase in intensity which comes back down as the rocket falls back into the atmosphere. And these curves compare the intensity as measured with the counter shielded with about 150 milligrams of aluminum with one not so shielded. In this way we obtained in the rather early days in this game about four or five years ago a rough idea of the energy and intensity of this effect in the fairly low altitudes in the auroral zone.

Figure 14 plots another flight made by Dr. McDonald of Iowa in '54.* This is a similar case with a very great increase in intensity. The peak altitude is about 100 kilometers

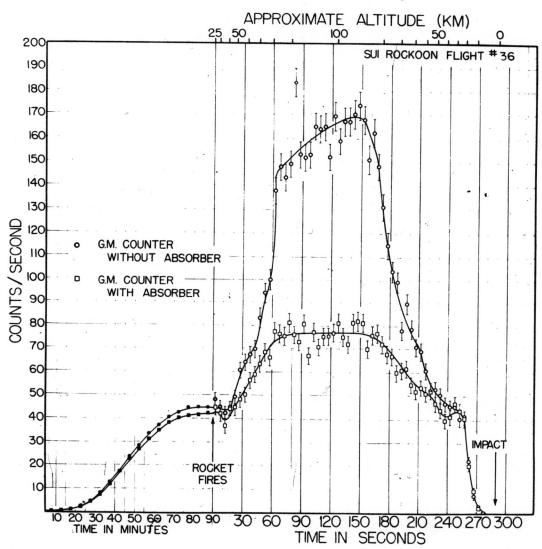

Figure 13

here. And here again (in the lower curve) is the absorption so that we have some idea of the hardness and the nature of the radiation.

Figure 15 shows the way in which this effect varies with geomagnetic latitude in a series of rockoon flights in the Arctic. And this (A) is the normal cosmic ray intensity,

and this great peak (B) is one roughly corresponding to the greatest incidence of visual aurorae.

Figure 16 is a sample balloon result. Dr. Winckler at Minnesota and Dr. Anderson at Iowa in balloon experiments have both observed the soft radiation now, at altitudes of about 110 thousand feet.** At bottom is a sample case of an ionization chamber, the rate has been going along for an hour or two, and then suddenly it has this big spike and

*McDonald's work is included in figures 13-15, from Van Allen, J.A., "Direct Detection of Auroral Radiation with Rocket Equipment," *Proc. N.A.S.*, *43*, 57-62, 1957.

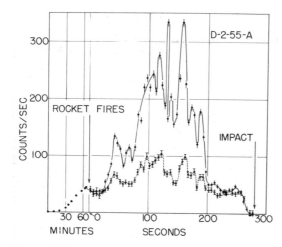

Figure 14

then other rather large excursions. The single Geiger counter (center) shows a similar effect, the ratio of these two effects giving one an idea of the energy of the radiation. A third detecting instrument used by Anderson was a three-fold cosmic ray telescope which showed (top) nothing like these large excursions in fractional magnitude. On the various combinations of evidence both from Winckler and Anderson's work this is reliably identified as X-radiation with energy about 100 kilovolts.

Figure 17 is another case of Dr. Anderson's, and a similar situation occurred here.

Finally I show a freshman physics slide (Figure 18) giving some idea what our tentative interpretation is. Here would be represented the earth with its magnetic field. We are about 2500 kilometers and we visualize this whole region from there on up as being more or less filled up with this plasma. Individually, electrons can go spiralling

**Anderson, K.A., "Occurrence of Soft Radiation During the Magnetic Storm of 29 August 1957," *J. Geophys. Res.*, *62*, 641-644, 1957.

Anderson, K.A., "Soft Radiation Events at High Altitude During the Magnetic Storm of August 29-30, 1957," *Phys. Rev.*, *111*, 1397-1405, 1958.

Winckler, J.R., and K.A. Anderson, "High-Altitude Cosmic-Ray Latitude Effect from 51° to 65° N Geomagnetic Latitude," *Phys. Rev.*, *108*, 148-154, 1957.

around this field and of course in fairly tight spirals, and go back and forth, and they can go as deep in the atmosphere as they can stand it before they run dead due to ionization. So presumably that is the gross explanation of the floor of the effect around 1000 kilometers. We don't of course know how far out this goes, although there are some sort of general energetic arguments which would suggest that it would not extend more than the order of a couple of earth radii or perhaps less. This space is filled with spiralling particles which we are now penetrating.

Let me write down some numbers.

$J_0 \sim 10^8$ to $10^9/\text{cm}^2\text{sec}$, if bremsstrahlung interpretation correct
$n \sim (0.1$ to $1.0)/\text{cm}^3$
$E \sim 40$ kev.
Energy Flux $\gtrsim 10$ ergs/cm^2sec

I'd like to say these are, of course, speculative. The observations we believe to be quite sound and reliable now. The speculation is in the mechanism and this is on the basis of the interpretation I suggested; namely, that what we are measuring actually in the counter is the bremsstrahlung formed by the electrons beating on the nose cone of the satellite. So if that interpretation is true, then we combine the theoretical calculation for bremsstrahlung generation with the Gamma ray or X-ray efficiency of the Geiger counter, which we measured experimen-

Figure 15

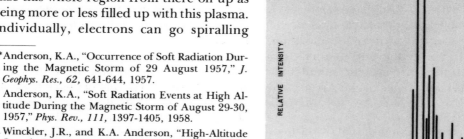

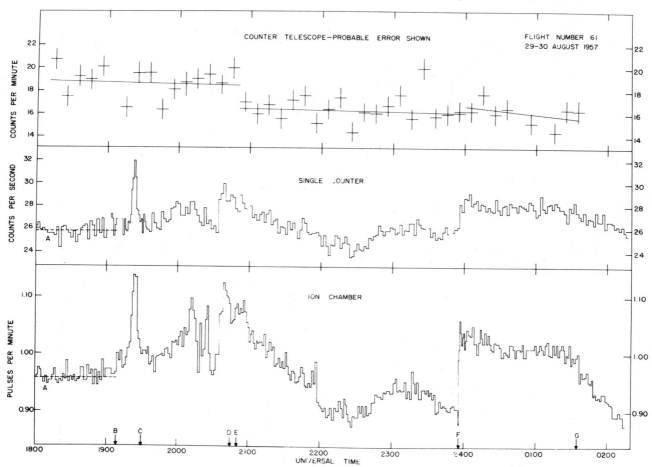

Figure 16

tally, and finally arrive at the fact that the flux of electrons, if this interpretation is true, is of the order of ten to the ninth per square centimeter per second; that the number density in space is of the order of a fraction to one per cubic centimeter at this energy; that the average energy is of the order of 40 kilovolts; and that the energy flux is of the order of ten ergs per square centimeter per second.

We think that these results may very likely be closely related to the general theoretical structure of Chapman and Ferraro* and we have already developed some interesting lines of speculation about the other conse-

quences and aspects of the situation. One appears to be that this intensity of radiation would likely result in a significant heating of the upper atmosphere and we already have some preliminary quantitative estimates of that. In the second place, of course, this plasma would be a source and certainly is a source, if it's really there as we believe it to be, of radio noise of calculable intensity. That so far doesn't appear to be an intensity

*Chapman, S., and J. Bartels, *Geomagnetism*, Vol. II, 853 ff., Oxford, 1940.
Chapman, S., and V.C.A. Ferraro, "A New Theory of Magnetic Storms," *Nature, 126,* 129-130, 1930; *Terr. Magn., 36,* 77-97, 171-186, 1931; *Terr. Magn., 37,* 147-156, 1932.

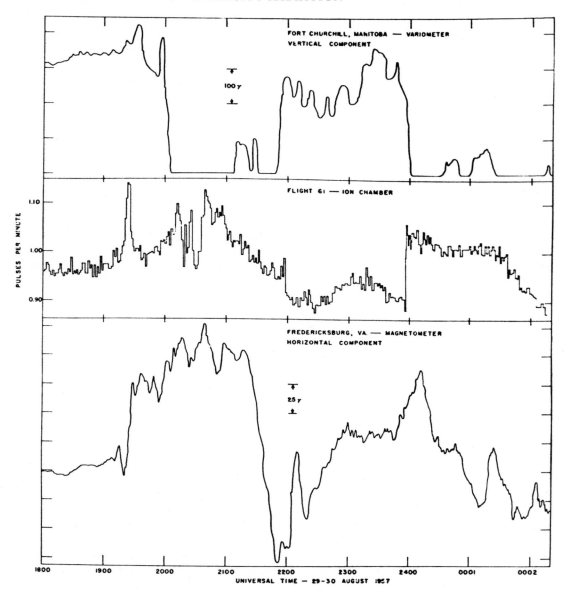

Figure 17

level which is observable from the earth through the ionosphere but should be observable from a vehicle above the ionosphere. And perhaps the most important significance, we believe, is the likely intimate connection between these results and the occurrence of visual aurorae. We suppose that this plasma is a semi-permanent feature of

the very high atmosphere that is being, say, continuously replenished from the sun, and that perturbations of it result in the formation of visible aurorae and that the particles are more or less continuously being fed down from above into lower altitudes where they run dead due to ionization in the atmosphere and that aurorae are, so to speak,

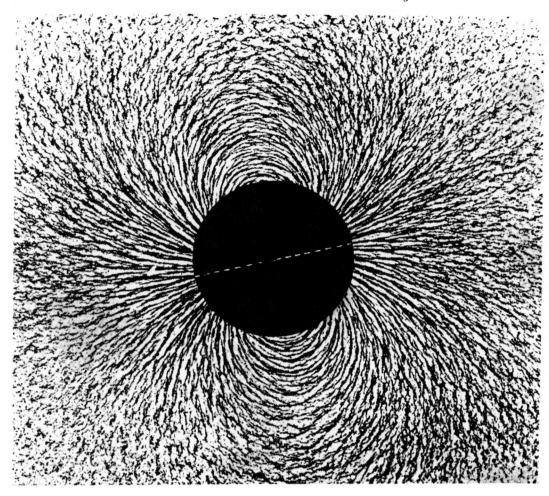

Figure 18

the leakage out of this reservoir plasma. This is, of course, speculative. In the next place, it may well be and likely is that some of the drift affects of these charged particles in the earth's general magnetic field may be responsible, may be the detailed mechanism, for geomagnetic storms.

I make only one final remark: it's that the radiation intensity above 1000 kilometers exceeds about 60 milliroentgens per hour. Now this was determined directly in the laboratory by exposing the 1958 Alpha apparatus to an X-ray beam of about 50 to 80 kilovolts energy and comparing its reading with an R-meter alongside whose reading we measured. This 60 mR per hour as a lower limit may give the biological flight people something to worry about. It would mean that a person would get a weekly tolerance dose in about two hours on the latest AEC standards, I believe. However, it's relatively easily shielded by a millimeter of lead which fairly well cuts it down to a low value.

Thanks very much.

Dr. Richard W. Porter: Would you please stay there for a few minutes? We are already 25 minutes late for the press conference in the other room, but I think the subject is of

73

sufficient importance that we certainly ought to have some discussion of it. So let's take about five or ten minutes at the most for questions.

Q: Is there any possibility that this radiation is caused by protons instead of electrons?

Dr. Van Allen: Well, of course, actually we don't know. The wall thickness of the counter excludes protons of energy less than 35 Mev, that's one point. It excludes electrons of energy less than three Mev and it's rather considerably absorptive for photons of energy less than 40 or 50 kilovolts. Now, if the auroral results of McIlwain do show protons present at the same time, they are of an order of magnitude lower flux and they are rather, and on the average, considerably lower in energy. So based on what we know, which of course is not conclusive, with respect to this situation, it appears very unlikely that what we are measuring is due to protons, but we do believe they are present but undetectable in our arrangement, and their bremsstrahlung, of course, would be of relatively negligible intensity.

Q: I'd like to make sure I understand the arguments for charged particles and against Gamma radiation: the idea is that the only mechanism you can conceive of by which the stuff is kept at high altitudes and kept from getting down to the lower ones is magnetic confinement rather than any form of absorption?

Dr. Van Allen: Yes, that's the essence of it. For example, if there were X-rays coming in from outer space or from the sun or some place, then there is no reason for them to stop at 100 kilometers because there is a negligible amount of material at, let's say, 100 kilometers so that we can't see anything to prevent them from coming right on down. The wall thickness of the counters is one-and-a-half grams so that the amount of material between 1000 and 100 kilometers is trivial in comparison. So we just think we just have to have the magnetic field holding them up.

Dr. W.W. Kellogg: I didn't get the number on the slide that you referred to as one of your earlier slides estimating the radiation due to just cosmic rays alone . . . (inaudible).

Dr. Van Allen: Yes.

Dr. Porter: What's the background cosmic ray count?

Dr. Van Allen: It's of the order of one-thousandth of this effect, let me see, . . . I am afraid I've forgotten that figure myself . . . it's, at any rate it's about a thousandth of this effect. It's . . .

Dr. Porter: 24 hours, I think . . .

Dr. Van Allen: It was on a 24-hour basis on that side, but anyway if you divide 60 by a thousand that's the right order of magnitude.

Question inaudible: . . . sharpness of the boundary . . .

Dr. Van Allen: Yes, that's taken from the question that has to do with the sharpness of the transition and the boundary between, say, 128 counts per second and a very much greater rate is of the order of a hundred kilometers thick in vertical altitude. Now that's deduced by taking the time of travel and the way that the satellite is moving. It's climbing fairly rapidly in that case at that time.

Q: If I'm not mistaken there was a solar flare. Is there any effect from a solar flare?

Dr. Van Allen: Yes, there was this very great geophysical event on the night of the 10th and morning of the 11th of February, about ten days after we had been up and we haven't finished the analysis of the data yet to pass on that point but we are very keenly interested in it.

Q: Some auroras and geomagnetic storms are supposed to be caused by variable emissions from the sun and you suggest that some auroras may be caused by the mechanism you mentioned. Are these different types of things both going on at the same time or . . .

Dr. Van Allen: Yes, the question is whether—Dr. Fritz asks whether the aurorae are generally supposed to be due to outbursts from the sun and how is that to be fitted in with what we are saying here. Well, of course, this again is, of course, speculative. But we suppose that there is this sort of big reservoir of plasma which is built up over a period of time from gas from the sun and then it slowly is worn away on the underside by loss in the atmosphere and that then from time to time it is replenished by a new burst from the sun and so its actual density may be a very rough function of time and not at all steady in responding to solar activity. So presumably the basic thing is the lifetime, how long it can remain there, because it's what we call sort of invisible radiation, that is, from the ground. As far as we can tell there's no way to tell it's there from the ground. So presumably an aurora would correspond to the arrival of a fresh batch of solar gas which would perturb the situation, replenish it and stir it up, so to speak, and then there would be a greater than ordinary leakage at the ends.

Q: I believe you mentioned that in respect to the average lifetime of the electron against stopping by ionization effects is in the order of hours?

Dr. Van Allen: Hours, at a thousand kilometers.

Q: Would it not be reasonable to expect some sort of diurnal effect?

Dr. Van Allen: You mean on the basis of the atmospheric density varying at . . .

Q: No, on the basis of the fact that the re-plenishment presumably has a diurnal effect.

Dr. Van Allen: Oh, we can't say anything about that observationally yet, but I think the general picture is the plasma gets kind of folded around the earth anyhow and may not be very directionally related to the direction of the sun, but I don't know.

Dr. Porter: Thank you very much, gentlemen. I can only express the hope that our future satellite program will be as illuminating in its respective fields of science as this apparently has in the field of radiation.

PUBLICATIONS AND FOLLOW-ON ARTICLES OF THIS WORK:

Van Allen, J.A., G.H. Ludwig, E.C. Ray, and C.E. McIlwain, "Observation of High Intensity Radiation by Satellites 1958 Alpha and Gamma," *Jet Propulsion,* pp. 588-592, September 1958.

Van Allen, J.A., "Radiation Belts Around the Earth," *Scientific American, 200,* 39-47, 1959.

Ludwig, G., "Cosmic Ray Instrumentation in the First U.S. Satellite," *Rev. Sci. Inst., 30,* 223-229, 1959.

Van Allen, J.A., "The Geomagnetically-Trapped Corpuscular Radiation," *J. Geophys. Res., 64,* 1683-1689, 1959.

Yoshida, S., G.H. Ludwig, and J.A. Van Allen, "Distribution of Trapped Radiation in the Geomagnetic Field," *J. Geophys. Res., 65,* 807-813, 1960.

Loftus, T.A., "Disturbance of the Inner Van Allen Belt as Observed by Explorer I," M.S. Thesis, University of Iowa, August 1969.

TOWARD
A
HISTORY

FROM BUMP TO CLUMP: THEORIES OF THE ORIGIN OF THE SOLAR SYSTEM 1900-1960

STEPHEN G. BRUSH

What is the scientific purpose of the space program? Among other things, it is to obtain evidence bearing on the origin of the solar system. As is well known from the history of science, the collection of empirical data is heavily influenced by hypotheses and theoretical preconceptions. Thus in order to understand the scientific history of the space program we must examine the cosmogonic theories that were dominant at its inception, and look at the interaction between theories and new discoveries resulting from the program itself. In this paper I will attempt to sketch the development of theories in the 20th century prior to 1960; this may help to provide the necessary background for studying the rich and complicated history of the

subsequent decades.

My account will focus on three men who had major impacts on modern theories of planetary formation: Thomas Chrowder Chamberlin (1843-1928), Henry Norris Russell (1877-1957), and Harold Clayton Urey (b. 1893). They represent the inputs of geology, astrophysics and chemistry, respectively, thereby highlighting the interdisciplinary nature of the subject. All three are American, a fact which is somewhat misleading since we will also recognize the essential contributions of Henri Poincaré (French, 1854-1912), James Jeans (British, 1877-1946), Harold Jeffreys (British, b. 1891), C.F. von Weizsäcker (German, b. 1912), Hannes Alfven (Swedish, b. 1908), Gerald Kuiper (Dutch-American, 1905-1973), and Otto Schmidt (Russian, 1891-1956).

This is not a success story. To be sure, we know much more about the solar system and its history now than we did 100 years ago. Yet we are *less* confident that we have the correct answer to the fundamental question: were the earth and other planets formed in the same process that produces all stars, or

Stephen G. Brush is Professor in the Department of History and the Institute for Physical Science and Technology at the University of Maryland. Author of numerous works in the history of modern physics, he has lately turned his attention to the history of geophysics and astrophysics.

78

does it take some special rare event to generate a planetary system? Until we can answer that question we cannot even estimate the order of magnitude of the probable number of other habitable planets in the galaxy, and thus we cannot say whether intelligent life is likely to be extremely rare or very common in the universe.

The origin of the solar system is the oldest unsolved problem in science (apart from questions about the origin and nature of life); leaving aside the ancient puzzle of the origin of the entire universe, we could say that ever since the acceptance of the Copernican heliocentric system early in the 17th century, the problem in its modern form has intrigued and frustrated some of the most brilliant scientists. The list of proposed solutions seems to include all conceivable possibilities; perhaps success will come eventually not by thinking up a completely new idea but by refuting all the objections to an old hypothesis.

I. Nebulae, Planetesimals, and The Big Bump

At the end of the 19th century most astronomers accepted Laplace's "nebular hypothesis," proposed 100 years earlier.[55] According to Laplace, the atmosphere of the primeval sun extended throughout the entire space now occupied by planetary orbits; it was a hot, luminous rotating cloud of gas, thought to be similar to the nebulae which the best telescopes of the time could not resolve into separate stars. As the nebula cooled it contracted; conservation of angular momentum then required it to rotate more rapidly. By hypothesis, the gas rotated like a rigid body in the sense that the angular velocity was the same at all distances from the center, so that the linear velocity would be greatest at the periphery. Eventually the centrifugal force on the outer portion of the nebula would exceed the gravitational attraction toward the center, and a ring of gas would separate and remain at the same distance while the inner part continued to con-

tract. By hypothesis, again, the gas in the ring would collect into a single large sphere, which would then gradually cool and condense to a liquid or solid planet. Meanwhile the contracting nebula would spin off additional rings at regular intervals, until finally only the sun was left at the center. Satellites could be formed by a similar process of ring separation as the protoplanetary sphere cooled down and condensed.

The nebular hypothesis was closely connected with 19th century geological theories, which generally presumed that the earth had been formed as a hot fluid ball and then cooled down, solidifying on the outside first.

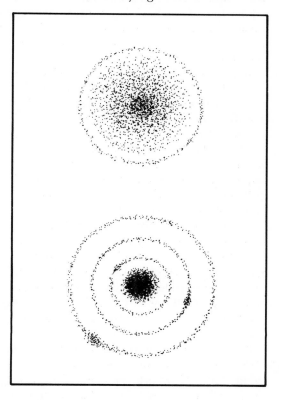

Figure. 1. The nebular hypothesis, adapted from J.P. Nichol, *Views of the Architecture of the Heavens* (Edinburgh: Tait, 1837), facing p. 170. The upper sketch shows initial separation of a ring at the outside of the rotating nebula; the lower sketch shows three separated rings, beginning to form planets, and a central condensation forming the sun.

According to the "contraction theory," the solid crust would not contract as rapidly as the fluid interior, so it would have to wrinkle in order to adjust its diameter to that of the shrinking core; in this way one could explain various geological formations.[33]

Lord Kelvin, the most influential British physicist of the 19th century, adopted the general scheme of a cooling earth but attacked two features that geologists relied on for their explanations. First, he estimated the time required to cool down from an initial hot fluid state and found it to be only 20 to 100 million years compared to the hundreds of millions of years geologists assumed to have been available for slow processes like erosion to produce their observed effects. Second, Kelvin showed that the fluid interior could not be so large as to extend to within 30 or 40 miles below the surface (as the geologists thought it should); in fact he went to the other extreme and concluded that the entire earth is now solid as "as rigid as steel."[10,15]

The debate between Kelvin and the geologists on the "age of the earth" was of course eventually settled by the discovery of radioactivity, which not only provided a source of heat to replace that which was lost from the original store (thereby invalidating Kelvin's conduction calculations) but also furnished a direct method for estimating the ages of some surface rocks. By 1905 Rutherford and his colleagues were proposing a time scale of billions of years, as part of the revolution that was sweeping through science.[11]

But even before the implications of radioactivity had been generally understood, T.C. Chamberlin challenged Kelvin's theory of the cooling earth in an assault so successful and far-reaching that it overthrew the nebular hypothesis itself.[17] Today it is hard to appreciate what a tremendous feat this was, because we now realize that the hypothesis had some serious flaws that had already been pointed out decades earlier. Yet no one had persuaded astronomers that

Figure 2. T.C. Chamberlin
Source: *Biographical Memoirs of the National Academy of Sciences*, volume 15 (1934).

patching up Laplace's theory was less fruitful than looking for a fundamentally different hypothesis; no one had shown geologists that the evidence for a hot primeval earth a few tens of millions of years ago was really quite flimsy; and no one had dared to tell Lord Kelvin that he was wrong in his basic assumptions about earth history. Chamberlin, a geologist who ventured into the apparently more difficult and prestigious field[9] of theoretical astronomy in his 50s, was hailed as the brash American who pulled the tail of the British lion and liberated geologists from the tyranny of the truncated time scale. At the same time he introduced into planetary cosmogony a hypothesis—accretion of cold solid particles—that, despite temporary rejection, has become an essential feature of most modern theories.[8,91]

Chamberlin's original objection to the nebular hypothesis was based, as one might expect, on geological considerations.[16] Having studied the glacial formations in North

America, he examined the contemporary attempts to explain the cause of the Ice Age, in particular the hypothesis that the earth originally had an atmosphere rich in carbon dioxide. A drop in the carbon dioxide content supposedly reduced the absorption of solar heat and thus lowered global temperatures. But when Chamberlin learned of calculations, based on the kinetic theory of gases, showing that gases at high temperatures would have molecular velocities great enough to escape the earth's gravitational field, he realized that the notion of a dense carbon dioxide atmosphere was inconsistent with the assumption that the earth had once been a hot fluid ball; not only carbon dioxide but all the other gases in the atmosphere would have escaped at temperatures high enough to melt rocks.

When Chamberlin looked into the possibility that the earth had been formed by accretion of cold solid particles, he found that this idea had indeed been discussed by astronomers under the name "meteoritic hypothesis." But it seemed to have a fatal defect: planets formed by combining solid particles moving in adjacent circular orbits would have retrograde rotation. The reason is that according to Kepler's third law, linear velocity decreases with distance from the sun, so the particle in the inner orbit would be moving faster than the one in the outer orbit just before they collided, and the combined body would have a net backwards rotation (Figure 4). Since it was thought that all planets (with the possible exception of Uranus and Neptune) have direct rotation, accretion from solid particles did not look very promising.

But the astronomers who rejected the meteoritic hypothesis on the basis of Kepler's third law had forgotten to apply Kepler's other two laws. In general the particles would move in elliptical orbits (first law), and a given particle would move faster in the part of its orbit which is closer to the sun (second law). Chamberlin showed by analyzing several examples that unions of particles

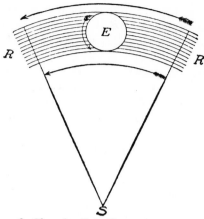

Figure 3. Chamberlin's illustration of the explanation of direct rotation on the nebular hypothesis: "RR represents a ring of gas moving as a unit and hence the outer portion the faster. If converted into a spheroid, E, centrally located, the rotation is forward, as shown by the arrow." *Origin of the Earth* (Chicago: University of Chicago Press, 1916), p. 91.

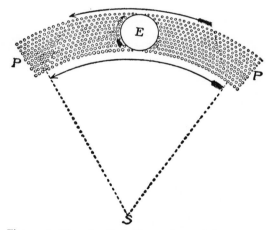

Figure 4. Chamberlin's illustration of the rotation produced by combining particles moving in adjacent circular orbits, with linear speeds decreasing with distance from the sun according to Kepler's third law. "PP represents a belt of planetesimals revolving concentrically about the center, S. If these collect about the central point of the belt into a spheroid, E, by the enlargement of the inner orbits or the reduction of the outer ones, the concentric arrangement remaining, the rotation will be retrograde, as shown by the arrow." T.C. Chamberlin, *The Origin of the Earth* (Chicago: University of Chicago Press, 1916), p. 92.

moving in intersecting elliptical orbits would be more likely to leave the resulting particle with direct rotation.[16,20,31] (Figures 5,6)

Here and elsewhere Chamberlin had the assistance of a young astronomer, Forest Ray Moulton (1872-1952), who was completing his Ph. D. research at the University of Chicago where Chamberlin headed the Geology Department. In addition to reviewing and working out the details of Chamberlin's ideas, Moulton put together the objections to the nebular hypothesis in a comprehensive paper published in 1900.[60] While his name became attached to Chamberlin's theory, Moulton's major contribution to cosmogony was to convince astronomers that the nebular hypothesis must be abandoned.

The major defect of the nebular hypothe-

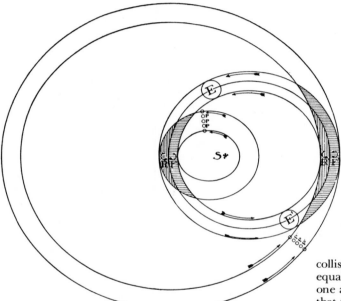

Figure 5. Chamberlin's explanation of the direct rotation of a planet produced by accretion of planetesimals in circular orbits onto a nucleus in an elliptical orbit:

A diagram intended to illustrate the proportion of normal cases of collision that will tend to give *forward* rotation as against *retrograde* rotation, a uniform distribution of planetesimals being assumed. To avoid unnecessary complication, a belt of planetesimals in the minimum normal orbits that permit collision, and another belt of those in the maximum normal orbits that permit collision, each having the breadth of the planetary nucleus, are chosen, as these are the cases of greatest differential velocities, and hence the most effective in producing rotation. The intermediate cases are not only less effective because of less differential velocity, but because the collisions of the two classes are more nearly equal in number and more nearly neutralize one another. In inspecting, let it be noted that the nucleus, E, at the left, is moving faster than the planetesimals in the area of possible collision, and hence that the planetesimals which it overtakes on its inner side tend to cause forward rotation, while those which it overtakes on its outer side tend to produce retrograde rotation. The collisions in the larger area, F, favoring forward rotation, are much more numerous than those in the small area, R, favoring retrograde rotation. At the right, the planetesimals are moving faster than the nucleus E', and those that overtake it on the outside in the area F' tend to forward rotation, while those that overtake it on the inner side in the area R' tend to retrograde rotation. In both cases, forward rotation is favored. If belts between these limiting belts be drawn, the difference between the two classes of areas will be less, but of the same phases.
Source: ref. 20, p. 76.

82

sis in 1900 was its failure to explain the distribution of angular momentum in the solar system. Laplace's spin-off of successive rings should have left the sun with much greater rotational speed than it now has. In fact Jupiter has most of the angular momentum of the solar system, contrary to what one would expect from any reasonable estimate based on the nebular hypothesis. On a smaller scale, the discovery in 1877 that Mars has a satellite (Phobos) which goes around it in only one-third of the rotation period of the planet contradicts the assumption that satellites have formed from rings spun off by the planet's nebula.

Chamberlin was primarily interested in

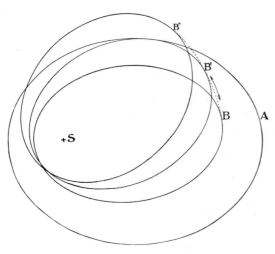

Figure 6. Chamberlin's explanation of why a perturbation of the elliptical orbits of planetesimals, causing them to intersect, will lead to accreting collisions more likely to produce a body with forward rotation.

Diagram showing that in the shifting of orbits, the first contingencies of collision favor forward rotation. *B* represents a smaller elliptical orbit within a larger one, *A*. If *B* be shifted progressively in the direction *B'*, *B''*, it will first come into possible collisional relations with *A* on its inner side, and at this point a body in the orbit *A* is moving faster than a body in the orbit *B*, as shown by the large orbit the former describes, and, the collision being on the inner side, forward rotation is favored. Source: ref. 20, p. 77.

the thermal and mechanical development of the earth rather than the rotation of the sun. He suggested that the accumulation of particles by the growing earth might have been so slow that the heat released by conversion of mechanical energy would be mostly dissipated into space and would never produce a large amount of melting.[17] Thus he agreed with Kelvin that the earth is now entirely solid, a view that prevailed until 1926 when Jeffreys established the existence of a liquid core.[14]

Nevertheless once Chamberlin had become interested in astronomical problems he could no longer confine himself to geology. Looking at James Keeler's photographs of spiral nebulae, he speculated that the two prominent arms belonged to two previously-distinct celestial objects. From this thought, and contemplation of solar prominences, he was led to the idea that a planetary system could be generated when another star passed close to the sun. He did not require an actual collision though this was being suggested by others.[8] The tidal force of the intruder would cancel the gravitional force holding in the solar gases on the near and far sides of the sun, allowing two filaments of material to flow out; the filaments would then be curved by the continued action of the intruder as it recedes (Figure 10). Chamberlin assumed that the filaments would eventually condense into small solid particles which would be captured into orbits around the sun.[18,19]

The Chamberlin-Moulton theory (published in 1905) thus consisted of two distinct hypotheses: (1) close encounters of two stars, drawing filaments of gaseous material out of one of them to form a spiral nebula; (2) condensation of the gases to small solid particles, called "planetesimals" (i.e. infinitesimal planets), which accrete to form planets and satellites.[19,61] When it later became clear that the spiral nebulae are galaxies rather than objects that could be as small as planetary systems, Chamberlin dropped this part of the first hypothesis but retained the as-

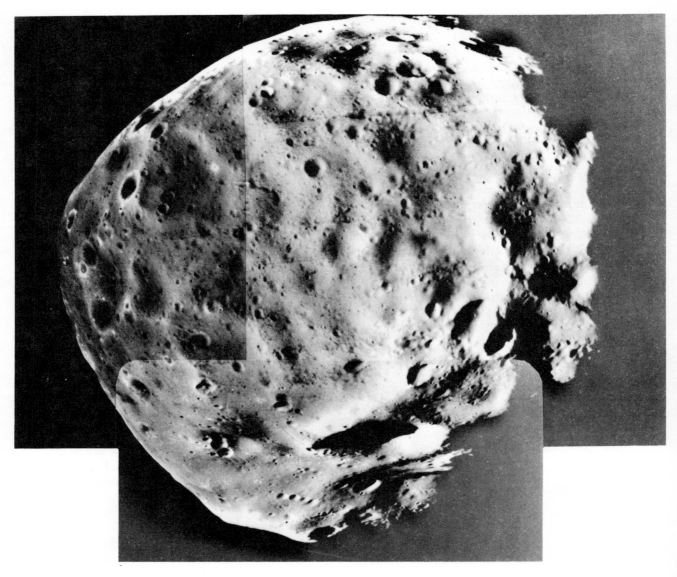

Figure 7. Phobos, the inner satellite of Mars, discovered by Asaph Hall in 1877. The fact that its period of revolution is only one-third of the period of rotation of the planet provided an argument against the nebular hypothesis. (Source: NASA)

sumption that two stars interacted in order to release into space the material from which planets formed.[21]

In 1916-17 Harold Jeffreys and James Jeans in England independently adopted the first hypothesis of the Chamberlin-Moulton theory, but rejected the second. Jeffreys argued that high-velocity collisions among the planetesimals would vaporize them so quickly that the material would remain gaseous until it collected into planets; thus he proposed to return to the 19th-century assumption that the earth was originally a hot fluid ball and has been cooling down.[45] Jeans was more interested in developing idealized mathematical models to represent the initial ejection of material from the sun under the tidal influence of the other star.[42] Thus Jeans concentrated on the astronomical side of the theory while Jeffreys developed it from a geophysical viewpoint.[46]

The tidal theory, whether the Chamberlin-Moulton or Jeans-Jeffreys version, was generally accepted by astronomers until 1935, even though it was never worked out in sufficient detail to provide a convincing explanation of the quantitative properties of the solar system. Its supporters believed that the tidal theory could overcome the major defect of the nebular hypothesis by showing at least qualitatively how most of the angular momentum could have been given to the major planets rather than to the sun. In the meantime, Henri Poincaré had demonstrated a theorem that seemed to provide another serious objection to the nebular hypothesis: if the present mass of the planets were spread out over the entire volume of the solar system, this material would be such a low density that it would dissipate into space before condensing.[66] The filaments postulated in the tidal theory would not have to be spread out over such a large volume, so this difficulty would be avoided.

Of course astronomers recognized that any theory which required the encounter of two stars to form planets would entail an extremely small frequency of planetary sys-

tems in the universe. This was consistent with the failure (before 1940) to find any convincing evidence for nonsolar planetary systems. Jeans at one time seemed to take perverse pleasure in the idea that we are the result of a chance event that has only happened once in the universe and (because the stars are decaying and thinning out by expansion) will probably never happen again.[43] Later he changed his mind and postulated that stars were much larger in the past so the frequency of collisions and hence of planetary systems was correspondingly larger.[44]

2. Astrophysics Strikes Back

In 1796 Laplace proposed his nebular hypothesis, not in one of his technical papers on celestial mechanics, but in a non-mathematical book on astronomy intended for the layman.[55] Similarly H.N. Russell in 1925 began to think about the origin of the solar system when working on a textbook, and presented his criticisms of the tidal theory in *Scientific American, Saturday Review,* and finally in a series of public lectures (1934), but he never discussed the subject in an astronomical journal.[8,71] Do astronomers still feel that cosmogony is not quite an appropriate subject for serious research, and does this attitude account for the slow rate of progress? (Cf. ref. 12, pp. 45-47.)

Russell found two major objections to the assumption that material extracted from the sun by a passing star would condense into the planets of the present solar system. First, theories of stellar structure developed by A.S. Eddington and others in the 1920s indicated that gases from the interior of the sun would be at such a high temperature—on the order of a million degrees—that they would dissipate into space before they could condense into planets.[71,78] Second, a simple dynamical calculation showed that it would be impossible for the tidal encounter to leave enough material with the necessary angular momentum in orbits at distances from the sun corresponding to the giant planets.[58,71]

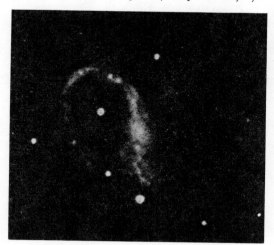

Figure 8. Chamberlin's example of the kind of nebula from which a planetary system might evolve: "A spiral nebula in which the two arms are especially distinct, HI 55 Pegasi." Source: T.C. Chamberlin, *The Origin of the Earth* (Chicago: University of Chicago Press, 1916), p. 125.

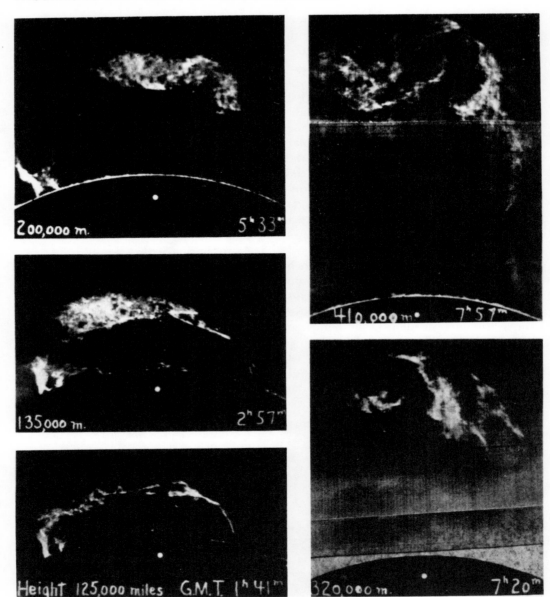

Figure 9. Chamberlin thought that solar prominences indicated the enormous tendency of solar gases to erupt into space, just barely restrained by gravity under ordinary conditions, and capable of ejecting a planet-forming filament when the sun's gravity is briefly neutralized by the tidal force of a passing star. These photographs show the rise of a very high prominence of May 29, 1919; the dot represents the size of the earth. Source: ref. 71, facing p. 101(photographs taken at Yerkes Observatory).

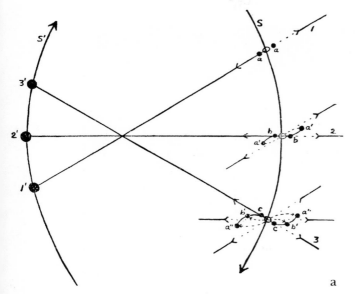

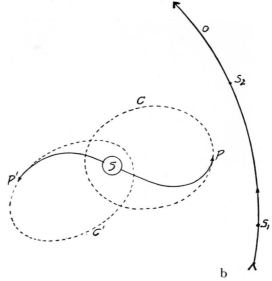

a

b

Figure 10 (a) Chamberlin's diagram showing the development of a spiral nebula from the action of intruder star S' on the sun S, at successive times 1, 2, 3. (b) Chamberlin's diagram showing development of orbits of portions of material in the filament left near the sun S, under the gravitational influence of the receding star at successive positions S_1, S_2. Source: T.C. Chamberlin, *The Origin of the Earth.*

Although R.A. Lyttleton, with Russell's encouragement, attempted to rescue the encounter theory by introducing a third star,[57] most astronomers seemed to think after 1935 that there was *no* satisfactory theory of the origin of the solar system: Russell had refuted the encounter theory, yet the fatal objections to the nebular hypothesis remained, and no other alternative seemed very plausible. Jeffreys in particular continued to insist that we simply have no adequate explanation for the existence of the solar system.[47]

Russell indirectly helped to demolish another argument that had previously been used by cosmogonists to account for the near-circularity of most planetary orbits. If the planets are initially formed with highly eccentric orbits, as seems likely for any type of encounter theory, one must find some mechanism to reduce the eccentricity. A popular choice was the hypothetical "resisting medium," that part of the dust and gas from the original nebula that did not condense into planets. Theoretically its viscous resistance should have helped elliptical orbits to evolve into more nearly circular ones. Russell was skeptical about this mechanism, pointing out that the medium would consist mainly of hydrogen and would be accreted

by the planets if they interacted with it; it would thus be hard to explain why the earth's atmosphere and oceans contain so little hydrogen.[72] Russell also encouraged H.P. Robertson at Princeton to look into the old claim by J.H. Poynting that the absorption and re-emission of solar radiation by small bodies in the solar system would decrease their angular momentum and eventually cause them to fall into the sun. There had been some dispute as to whether Poynting's result, derived from the ether theory in 1903,[67] was consistent with Einstein's theory of relativity. Robertson showed that there is indeed a dragging effect (though Poynting's formula is not accurate) and that particles less than one centimeter in radius in the vicinity of the earth's orbit would be swept into the sun in less than 40 million years.[69] Thus the "Poynting-Robertson effect"

87

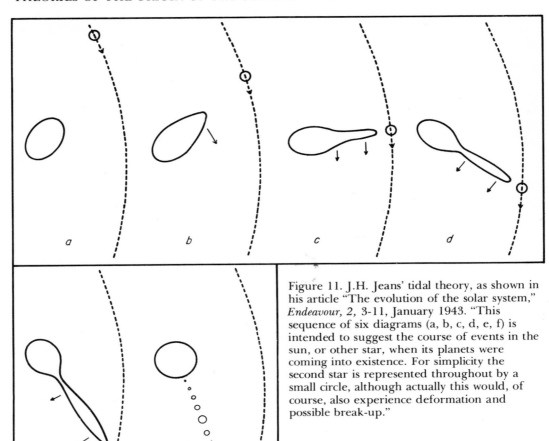

Figure 11. J.H. Jeans' tidal theory, as shown in his article "The evolution of the solar system," *Endeavour, 2*, 3-11, January 1943. "This sequence of six diagrams (a, b, c, d, e, f) is intended to suggest the course of events in the sun, or other star, when its planets were coming into existence. For simplicity the second star is represented throughout by a small circle, although actually this would, of course, also experience deformation and possible break-up."

makes it unwise to invoke a resisting medium to round up planetary orbits except under carefully defined conditions.

In addition to knocking out the encounter theories, Russell also made an important discovery which later removed one of the objections to the nebular hypothesis and substantially influenced its modern form. In the 1920s astronomers believed that the sun has roughly the same chemical composition as the earth; this would be consistent with the hypothesis that the earth was formed from material drawn out of the sun by a passing star. Thus the sun should contain substantial amounts of elements such as iron, silicon and oxygen but relatively little hydrogen and helium. Russell showed in 1929 that hydro-

gen is by far the most abundant element in the sun's atmosphere,[70] and other astrophysicists in the 1930s established that the same is true for many other stars and probably the universe as a whole.[82] If one assumes that the earth was formed from a cloud of material characterized by the typical "cosmic abundance" of elements, then most of the hydrogen originally present in this cloud— now called the "solar nebula"—must have been lost. Hence its original mass must have been much greater than that of the present planets, and its density could have been great enough to satisfy Poincaré's criterion for condensation.[66] Of course one still has to deal with the other long-standing objections to the nebular hypothesis, and in addition explain the chemical processes that produced planets with compositions radically different from the cosmic abundance table.

Since Russell's discovery of the predominance of hydrogen in the solar atmosphere ultimately transformed the character of planetary cosmogony by requiring the introduction of chemical considerations, just as it had a major impact on theories of stellar structure and evolution, it is worth a short digression to explain its background. How does one determine the chemical composition of the sun?

In the 1920s the primary source of information about the sun was the analysis of spectral lines in the light it sends us. The most important astronomical discovery of the second half of the 19th century[12] was the demonstration in 1859 by Kirchhoff and Bunsen that the characteristic frequencies of dark lines in the solar spectrum could be matched up with bright lines in the spectrum of known chemical substances determined in the laboratory.[50] The atoms in the solar atmosphere apparently absorb light coming from the hotter interior at the same frequencies they could emit light if vaporized by a Bunsen burner. Thus by detailed examination of the solar spectrum and comparison with the spectra of known elements, one can determine what elements are present in the sun's atmosphere, or at least that part of it known as the "reversing layer."

One might also expect that the intensity of a dark line is related to the abundance of the element responsible for it, but the relation is far from simple. A reasonably complete understanding of the situation had to await the development of the quantum theory and the theory that gases in the sun are in various states of ionization depending on temperature and pressure. The theory of ionization equilibrium, proposed by the Indian physicist M.N. Saha in 1920,[74] involves Planck's constant and the energy required to remove electrons from atoms ("ionization potentials") as parameters. From a comparison of the relative intensities of spectral lines identified with different ions, one can infer the temperature and pressure of the region in which the spectral lines are produced. [27,64] Further analysis allows one to estimate the relative abundances of these ions. This was first attempted by Cecilia Payne at Harvard in 1925, but she could not believe the result: the abundances of hydrogen and helium came out so much higher than expected that she assumed there was some hidden source of error in the method.[64]

Russell made a thorough investigation of all available data on the solar spectrum, and used the most recent theoretical estimates for the relative intensities of spectral lines produced by the same ion based on quantum theory. (Incidentally, he had himself contributed to atomic theory with his interpretation of spectral lines based on what is now called "Russell-Saunders" coupling.) He concluded that the high abundance of hydrogen estimated from ionization theory was correct, and that one could resolve some other difficulties in the theory of stellar structure by assuming that the gases in the interior of a star have a much smaller average molecular weight than previously thought.[70] The Danish astronomer Bengt Strömgren showed in the 1930s that most stars have a large proportion of hydrogen, at least in their early stages.[81] The Bethe-

Figure 12. Henry Norris Russell. From AIP Niels Bohr Library, Margaret Russell Edmondson collection.

Weizsäcker theory of stellar energy generation was based on the assumption that hydrogen atoms are fused into helium and heavier elements as the star gets older.[5,93] George Gamow's "big bang" cosmology postulated that all the matter in the universe was created in the form of hydrogen (or an equivalent mixture of protons and electrons), some of which is later turned into heavier elements.[4,29,30]

3. Intermission

In the decade following Russell's refutation of the encounter theory (1935), no single theory was supported by more than a handful of astronomers. Nevertheless there were some significant developments in this decade that influenced later work: (a) revival of the planetesimal hypothesis; (b) the concept of "magnetic braking" of the sun's rotation; (c) suggestions that the sun had encountered an interstellar cloud and captured from it the material that later formed planets; (d) claims for discovery of extrasolar planetary systems; (e) research on cosmic abundances of the elements. I will summarize these briefly.

(a) The Swedish astronomer Bertil Lindblad (1895-1965) showed that partly-inelastic collisions between particles initially moving with different speeds in eccentric orbits with different inclinations will tend to make all the particles move at similar velocities in circular orbits lying in a flat ring. Collisions between the particles would then occur with small relative velocities, thereby avoiding Jeffrey's argument that collisions would vaporize the particles. Lindblad suggested that a cold particle immersed in a hot gas would tend to grow by condensing the gas on its surface.[56] Dirk ter Haar in Leiden elaborated this idea by using the Becker-Döring kinetic theory of formation of drops in a saturated vapor, and reinforced Lindblad's proposal that solid particles could grow initially by non-gravitational forces.[34]

Jeffreys himself began to reconsider his objection to the planetesimal hypothesis and suggested that the vapour pressure of solids at very low temperatures might be below the pressure in the surrounding medium, so that condensation would outweight the vaporizing effect of collisions.[48] Alfred Parson published an estimate of the vapor pressure of iron ($10^{-46.7}$ atm at 273°K) which indicated that condensation would be favored in interstellar space,[63] and Jeffreys admitted in 1948 that his original objection to the planetesimal theory had thereby been answered.[49]

Fred Whipple proposed in 1942 that radiation pressure acting on particles in a dust cloud would tend to push them together; each of a pair of nearby particles would shield the other from the radiation, leaving an effective attraction.[79,96] This mechanism is similar to the "kinetic" explanation of gravity proposed by LeSage and others in the 18th and 19th centuries.[7] Whipple proposed condensation of dust particles initially as a means of star formation from the dark clouds studied by Bart Bok, but also used it as an initial stage in the formation of planetary systems.[97]

(b) Hannes Alfvén, whose early papers on cosmogony were communicated for publication by Lindblad, incorporated the concept of planetesimal accretion into his own theory but also added an important idea which removed a major objection to the nebular hypothesis. He showed that an ionized gas surrounding a rotating magnetized sphere will acquire rotation and thereby slow down the rotation of the sphere.[1] Ferraro had obtained this result earlier but did not suggest its possible use in cosmogony.[26] Alfvén proposed that the early sun had a strong magnetic field, and that its radiation ionized a cloud of dust and gas, which then trapped the magnetic field lines and acquired most of the sun's original angular momentum.[2,3] This mechanism of "magnetic braking" was later adopted by other theorists who rejected the rest of Alfvén's cosmogony.

If a star is rotating, the Doppler effect will cause a shift in the frequencies of spectral lines for radiation emitted by those parts momentarily moving toward or away from us, and by careful analysis of stellar spectra it is possible to estimate the speeds of rotation. Otto Struve and others found that stars in later stages of evolution generally rotate more slowly than those in earlier stages. There is apparently some fairly universal process by which a star loses most of its angular momentum at a particular stage of its evolution.[83] Whether or not this process involves the formation of planets, at least one can no longer use the slow rotation of the sun as an argument against the nebular hypothesis.

(c) Alfvén proposed that the sun encountered a cloud of neutral gas which became more or less completely ionized at the distance of Jupiter; the magnetic field of the sun prevented it from moving any closer. Another cloud, consisting of dust particles, was postulated to account for the terrestrial planets.[2] At about the same time (1944) Otto Schmidt in the USSR proposed that the sun had captured an interstellar cloud of meteorites; his theory was based on gravitational capture rather than electromagnetic effects. Like Alfvén he assumed that the earth was formed by accretion of cold solid particles.[75] This assumption provides a common basis for discussion of questions about the thermal history of the earth, evolution of its core, etc., for scientists who may disagree on whether the sun itself was formed from this cloud or encountered it later.

(d) In 1943, two reports of extrasolar planetary systems provided a new argument against all theories that treated the origin of the solar system as an extremely rare event. Both reports inferred the existence of an invisible third component, with mass much less than the smallest known stellar mass, from observations of a binary system.[68,80] The headline of Russell's monthly column in *Scientific American* proclaimed: "Anthropocentrism's demise: New discoveries lead to the probability that there are thousands of inhabited planets in our galaxy."[73] The obituary turned out to be premature as the validity of such "discoveries" became a matter of controversy; yet many scientists wanted to find life elsewhere in the universe and would not be happy with a theory that made it unlikely.[39,92]

(e) Research on the cosmic abundance of chemical elements, mentioned above in connection with Russell's interpretation of the solar spectrum, was also pursued through analysis of meteorites; much of this information was synthesized in a classic paper by V. Goldschmidt in 1937, and brought up to date in a review by Harrison Brown in 1949.[6,32]

One way to resolve the discrepancy between the high cosmic abundance of hydrogen and its low abundance at the earth's surface was to postulate that the earth's core contains a large amount of hydrogen. This was the proposal of Kuhn and Rittmann.[51] They criticized the standard iron core model of the earth on the grounds that there was no plausible mechanism for separating the iron during the evolution of the earth. Arnold Eucken, in response to this challenge, proposed a process by which the earth could have been formed from a hot gas of solar composition, losing hydrogen while concentrating iron in the core.[25]

4. A Nebula That Clumps and Coughs?

The postwar revival of the nebular hypothesis is mainly due to a paper by C.F. von Weizsäcker, published in 1944.[94] Weizsäcker postulated a gaseous envelope surrounding the sun and associated with its formation; in order to contain enough heavy elements to form the planets and at the same time have the high proportion of hydrogen and helium characteristic of the sun, this envelope must have had about one tenth of the mass of the sun. If the envelope or solar nebula were concentrated in a flat disk with diameter approximating that of the orbit of Pluto, its density would be relatively high (about 10^{-9} g/cm^3) and it might stay together long enough to develop a regular pattern of motions.

Whereas Laplace had assumed, rather implausibly, that the gaseous nebula would rotate like a rigid solid, Weizsäcker pointed out that there would be a tendency toward differential rotation with faster motion inside and slower outside, as in Kepler orbits. But friction between adjacent streams would tend to equalize their speeds by accelerating the outer stream and decelerating the inner one. This creates an instability, causing the outer stream to move further out and the inner stream to move inwards, resulting in turbulent convection currents and eventually the formation of a pattern of vortex motions. Each vortex moves in a circular orbit around the sun, and there must be an integer number of vortices in a ring. If one postulates exactly 5 vortices per ring Figure 13 then the nth ring will be at a distance $r_n = r_0 \varepsilon^n$ where ε is 1.9, giving an approximation to Bode's law of planetary distances. Weizsäcker assumed that the best place to

accumulate particles into planets would be the regions where adjacent vortices come into contact producing violent turbulence.

Weizsäcker's theory was initially greeted with enthusiasm, especially in the United States where it was reviewed by George Gamow and J.A. Hynek,[28] and by S. Chandrasekhar.[22] Ter Haar adopted it as a basis for further work, incorporating his own mechanism for condensing dust particles.[35] But subsequent work on turbulence theory by Heisenberg,[38] Weizsäcker,[95] and Chandrasekhar[23,24] indicated that the regular pattern of vortices originally postulated by Weizsäcker could not occur, but instead must be replaced by a range of eddy sizes.

The next major development was due to Gerard Kuiper, who envisaged planetary formation as a special case of a general process that ordinarily leads to binary stars.[52,53] He rejected Weizsäcker's assumption that planets would be formed in the region between vortices, postulating instead that the nebula would first break up into giant protoplanets by gravitational action. The protoplanets destined to form terrestrial planets would lose their hydrogen before condensing. Kuiper adopted Alfvén's magnetic braking mechanism to explain the present slow rotation of the sun.[54]

Harold Urey, who received the Nobel Prize in 1934 for his discovery of deuterium and was an influential member of the postwar "atomic scientists" community, became interested in the formation of the earth when he agreed to give a course on "Chemistry in Nature" with Harrison Brown at the University of Chicago in 1948 or 1949.[87] To prepare his first lecture on the heat balance of the earth he read Louis B. Slichter's 1941 article[76] and was surprised to learn that the temperature of the earth might actually be rising rather than falling. He wrote in 1952: "This led on to consideration of the curious fractionation of elements which must have occurred during the formation of the earth. One fascinating subject after another came to my attention, and for two years I have

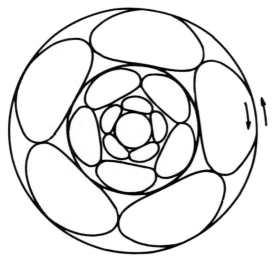

Figure 13. Pattern of motions in the solar nebula according to Weizsäcker's theory: "System of convection vortices around the sun. The outer arrow indicates the total rotation of the gas envelope, the inner arrow shows the sense of rotation of the vortices. The sun should be imagined as lying in a common center." Source: ref. 94.

Figure 14. Harold C. Urey.

later. Thus initially the cold earth had a core of "moon-like material" surrounded by a layer of silicates and iron, with oceans covering the entire surface. Later as the earth warmed up because of radioactivity, the iron flowed to the center (in spite of Kuhn and Rittmann's doubts); this process would release gravitational energy in the form of heat, which in turn would lead to convection in the mantle and the formation of continental land masses and mountains by the rising low-density material displaced from the core.

In an editorial three days later the *New York Times* rather ambiguously applauded Urey's theory as a romantic dream, the "folklore of a scientific age which helped dispel the image of scientists as dullards concerned only with facts. . . . Let's have more fiction of the type that Dr. Urey has given us. There is something epic about it."[62]

The reaction of prominent American scientists was favorable; Slichter organized a conference under the sponsorship of the National Academy of Sciences at Rancho Santa Fe, California, in January 1950 to discuss Urey's theory.[77] The group of 24 scientists included seven chemists, who favored a cold origin for the earth in order to account for the presence of water and the relative absence of noble gases. Geologists also supported a cold origin, arguing that global temperatures high enough to melt the crust would have produced more complete layering of the crust and mantle than is observed.

Urey published the first detailed account of his theory in a long article in 1951,[86] followed by a book in 1952.[87] He accepted Kuiper's modification of Weizsäcker's theory; he had learned about Eucken's 1944 theory only after his own ideas had been developed, probably from H.E. Suess who spent a year (1950-51) in Urey's laboratory at Chicago. While he found Eucken's theory "most impressive" and perhaps valid for the major planets, he could not accept the high temperatures which Eucken postulated for the formation of the earth. Rather than start-

thought about questions relating to the origin of the earth for an appreciable portion of my waking hours, and have found the subject one of the most interesting that has ever occupied me." (ref. 87, p. ix)

Urey presented his first ideas on the formation of the earth at a meeting of the National Academy of Sciences on October 26, 1949.[85] His most important conclusion, according to a report in the *New York Times,* was that the earth is heating up and the day is growing longer.[65] Urey assumed that the earth condensed from a dust cloud which contained a considerable amount of water vapor. Low density material condensed first and higher density material, especially iron,

ing with an initial separation of an iron core in the gaseous protoearth as Eucken proposed, Urey assumed that the terrestrial planets accumulated from planetesimals and initially consisted of a grossly homogeneous mixture of silicate and iron phases. The iron would initially be in an oxidized condition in the presence of cosmic proportions of water vapor; it was therefore necessary to postulate a later high-temperature stage during which the iron was reduced and partially fractionated from the silicates.

In his 1951 paper Urey emphasized the significance of the moon's present surface features as a record of conditions in an earlier stage of the history of the solar system. "Markings on the moon's surface indicate that iron-nickel alloy objects a few kilometers in radius fell on the moon.... The inference is that such objects also fell on the earth at this stage" (ref. 86, p. 212.) But traces of them were obliterated during the earth's geological development. The bombardment suddenly stopped and no major changes occurred since the moon became rigid, about four billion years ago (as estimated by Jeffreys). The moon "represents more nearly the composition of the original dust-cloud relative to the non-volatile elements than does the earth." (ref. 86, p. 241) The fact that the moon does not have the shape corresponding to isostatic equilibrium indicates that it was frozen a long time ago. Thus Urey prepared the way for the argument that exploration of the moon can give us information about the early history of the earth which cannot be found by studying the earth itself.

Urey's theory also had implications for the origin of life: chemical conditions in the early atmosphere should have been reducing rather than oxidizing. Urey suggested that laboratory experiments should be done on the production of organic compounds from water and methane in the presence of ultraviolet light and electric discharges;[88] the result of this suggestion was the famous experiment by Urey's graduate student, Stan-

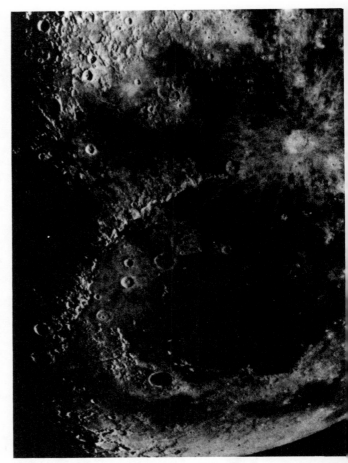

Figure 15. Mare Imbrium region of the moon. (a) Photograph taken Sept. 14, 1919 at Mt. Wilson Observatory.

ley L. Miller, which began the modern phase of research on the chemical evolution of life.[59]

In 1954 Urey indicated his doubts about the role of Kuiper's large protoplanets, on the grounds that it was difficult to understand how silicates could have evaporated from them to the extent necessary to explain the composition of the terrestrial planets while still retaining some water, nitrogen, and carbon.[89] Two years later he abandoned the hypothesis that protoplanets (in the sense of large masses of gas and dust of solar composition) had been involved in the formation of the terrestrial planets. Instead he

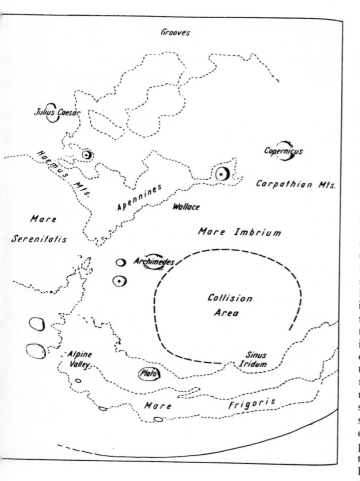

Figure 15. (b) Urey's interpretation of features in the Mare Imbrium region: "a large object striking near Sinus Iridum at a low angle partially rebounded from the surface, spreading its substance through approximately 180° between the eastern Carpathians and the western border of Plato. Immediately beyond the point of contact it dropped materials into Mare Imbrium, whose lavas were still to cover its floor. Just beyond the shores of the mare it dropped the Alps, Caucasus, Apennines and Carpathian mountains in radiating ridges, and in the region of the Haemus mountains, Mare Vaporum and Mare Nubium further long and short ridges. It also supplied the objects which ploughed out the grooves at still greater distances. These radial patterns are of two distinct kinds and must have been made by materials of different physical properties. The straight narrow grooves suggest materials of high density and tensile strength such as iron-nickel alloy ploughing through the surface while the ridges suggest silicates of low tensile strength. The Alpine valley also indicates that part of the meteoric body contained a high density body embedded in the silicate, which because of its greater momentum and energy per unit volume moved on through the region in which the silicates spread themselves into Mare Frigoris, sinking in its lavas or being subsequently covered by them. Sinus Iridum may have been produced by the planetesimal and may record the point of collision. Its width from Pr. La Place to Pr. Heraclides is about 230 km and this is approximately the correct dimension of the planetesimal. Source: ref. 86, pp. 221-26.

postulated that two sets of objects of asteroidal and lunar size, called "primary" and "secondary" objects, were accumulated and destroyed during the history of the solar system. The primary objects were suddenly heated to the melting point of silicates and iron, perhaps by explosions involving free radicals triggered by solar-particle radiation. After cooling for a few million years these primary objects "were broken into fragments of less than centimeter and millimeter sizes. The secondary objects accumulated from these about 4.3×10^9 years ago, and they were at least of asteroidal size. These objects were broken up ... and the fragments are the meteorites." (Ref. 90, p. 623) The reason for constructing this scheme was

to explain the presence of diamonds (presumably formed only at very high pressures) in meteorites.

At the first "Symposium on the Exploration of Space" (April 1959), Urey suggested that "the moon may be one of these primary objects, as I realized after devising what seemed to me a reasonable model for the *grandparents* of meteorites." (Ref. 91, p. 1727) As the *New York Times* headlined one of his speeches two years later, "Urey holds moon predated earth" and is one of the few relics of an early stage of the solar system."[84] Urey could therefore prescribe a set of chemical and physical observations to be made from the moon's surface to give information not only about meteorites but also the formation of the planets. He concluded his 1959 paper with the remark: "It is hoped that such observations will be forthcoming during the immediate years ahead." Urey's theory was not the only one available at the beginning of the space program, but it was certainly one that could generate many testable predictions and thus played an important role in the early days of lunar exploration.

Acknowledgment

This paper is based on research supported by the History and Philosophy of Science Program of the National Science Foundation.

REFERENCES

1. Alfvén, Hannes. "Remarks on the rotation of a magnetized sphere, with application to solar rotation," *Arkiv för Matematik, Astronomi och Fysik, 28,* No. 6, 1942.

2. Alfvén, Hannes. "On the cosmogony of the solar system," *Stockholms Observatoriums Annaler, 14,* No. 2, 1942, No. 5, 1943, No. 9, 1946.

3. Alfvén, Hannes. *On the Origin of the Solar System.* (Oxford: Clarendon Press, 1954).

4. Alpher, R.A., H. Bethe and G. Gamow. "The origin of chemical elements." *Physical Review, series 2, 73,* 803-4, 1948. See ref. 29, p. 398, note.

5. Bethe, H.A. and C.L. Critchfield. "The formation of deuterons by proton combination," *Physical Review, series 2, 54,* 248-54, 860, 1938.

6. Brown, Harrison. "A table of relative abundances of nuclear species," *Review of Modern Physics, 21,* 625-34, 1949.

7. Brush, Stephen G. *The Kind of Motion We Call Heat: A History of the Kinetic Theory of Gases in the 19th Century* (Amsterdam: North-Holland Publishing Company, 1976), pp. 21, 22, 48, 135, 336, 388, 396.

8. Brush, Stephen G. "A geologist among astronomers: The rise and fall of the Chamberlin-Moulton cosmogony," *Journal for the History of Astronomy, 9,* 1-41, 77-104, 1978.

9. Brush, Stephen G. "Planetary science: From underground to underdog," *Scientia, 113,* 771-787, 1978.

10. Brush, Stephen G. "Nineteenth-century debates about the inside of the Earth: Solid, liquid or gas?" *Annals of Science, 36,* 225-54, 1979.

11. Brush, Stephen G. "Scientific revolutionaries of 1905: Einstein, Rutherford, Chamberlin, Wilson, Stevens, Binet, Freud," in *Rutherford and Physics at the Turn of the Century,* edited by Mario Bunge and William R. Shea (New York: Science History Publications, 1979), pp. 140-71.

12. Brush, Stephen G. "Looking up: The rise of astronomy in America," *American Studies, 20,* No. 2, 41-67, 1979.

13. Brush, Stephen G. "Poincaré and cosmic evolution," *Physics Today, 33,* No. 3, 42-49, 1980.

14. Brush, Stephen G. "Discovery of the Earth's Core," *American Journal of Physics, 48,* 705-724, 1980.

15. Burchfield, Joe D. *Lord Kelvin and the Age of the Earth* (New York: Science History Publications, 1975).

16. Chamberlin, T.C. "A group of hypotheses bearing on climatic change," *Journal of Geology, 5,* 653-83, 1897. See esp. pp. 668-69.

17. Chamberlin, T.C. "On Lord Kelvin's address on the age of the earth as an abode fitted for life," *Science, 9,* 889-901, 1899; *10,* 11-18, 1899.

18. Chamberlin, T.C. "On the possible function of disruptive approach in the formation of meteorites, comets, and nebulae," *Journal of Geology, 9,* 369-92, 1901; *Astrophysical Journal, 14,* 17-40, 1901.

19. Chamberlin, T.C. "Fundamental problems of geology," *Carnegie Institution Year Book No. 3 for 1904,* (Washington, D.C.: Carnegie Institution, 1905) pp. 195-234; Ibid., *Year Book No. 4 for 1905,* pp. 171-85 (pub. 1906).

20. Chamberlin, T.C. and R.D. Salisbury. *Geology*. (New York: Holt, 1906), Vol. II, p. 76.

21. Chamberlin, T.C. *The Origin of the Earth* (Chicago: University of Chicago Press, 1916); *The Two Solar Families: The Sun's Children* (Chicago: University of Chicago Press, 1928).

22. Chandrasekhar, S. "On a new theory of Weizsäcker on the origin of the solar system," *Reviews of Modern Physics, 18,* 94-102, 1946.

23. Chandrasekhar, S. "On Heisenberg's elementary theory of turbulence," *Proceedings of the Royal Society of London, A200,* 20-33, 1949.

24. Chandrasekhar, S. and D. ter Haar. "The scale of turbulence in a differentially rotating gaseous medium," *Astrophysical Journal, 111,* 187-90, 1950.

25. Eucken, A. "Physikalisch-chemische Betrachtungen über die früheste Entwicklungsgeschichte der Erde," *Nachrichten von der Akademie der Wissenschaften in Göttingen, Mathematisch-physikalische Klasse* 1-25, 1944; "Über den Zustand des Erdinnern," *Naturwissenschaften, 32,* 112-21, 1944.

26. Ferraro, V.C.A. "The non-uniform rotation of the sun and its magnetic field," *Monthly Notices of the Royal Astronomical Society, 97,* 458-72, 1937.

27. Fowler, R.H. and E.A. Milne. "The intensities of absorption lines in stellar spectra, and the temperatures and pressures in the reversing layers of stars," *Monthly Notices of the Royal Astronomical Society, 83,* 403-24, 1923; "The maxima of absorption lines in stellar spectra (Second Paper), *Ibid., 84,* (1924), 499-515.

28. Gamow, George, and J.A. Hynek. "A new theory by C.F. von Weizsäcker on the origin of the planetary system," *Astrophysical Journal, 101,* 249-54, 1945. See also ref. 30, pp. 1, 103-10.

29. Gamow, George. "The origin and evolution of the universe," *American Scientist, 39,* 393-406, 1951.

30. Gamow, George. *The Creation of the Universe* (New York: Viking, 1952).

31. Giuli, R.T. "On the rotation of the earth produced by gravitational accretion of particles," *Icarus, 8,* 301-323, 1968. See also A.W. Harris, paper and letter quoted in ref. 8, p. 34, note 62.

32. Goldschmidt, V.M. "Geochemische Verteilungsgestze der Elemente IX," *Skrifter av Det Norske Videnskaps-Akademi i Oslo, Mat.-Naturv. Klasse* No. 4, 1937.

33. Greene, Mott. T. *Major Developments in geotectonic theory between 1800 and 1912*. Ph. D. Dissertation, University of Washington, 1978.

34. Haar, D. ter. "On the origin of smoke particles in the interstellar gas," *Bulletin of the Astronomical Institutes of the Netherlands, 10,* 1-8, 1948; *Astrophysical Journal, 100,* 288-99, 1944.

35. Haar, D. ter. "Studies on the origin of the solar system," *Det Kgl. Danske Videnskaberne Selskab, Matematisk-Fysiske Meddelelser, 25,* Nr. 3, 1948.

36. Haar, D. ter. "Recent theories about the origin of the solar system," *Science, 107,* 405-11, 1948.

37. Haar, D. ter, and A.G.W. Cameron. "Historical review of theories of the origin of the solar system," in *Origin of the Solar System,* edited by R. Jastrow and A.G.W. Cameron. (New York: Academic Press, 1963), pp. 1-37.

38. Heisenberg, W. "Zur statistischen Theorie der Turbulenz," *Zeitschrift für Physik, 124,* 628-57, 1948; "On the theory of statistical and isotropic turbulence," *Proceedings of the Royal Society of London, A195,* 402-6, 1948.

39. Huang, Su-Shu. "Extrasolar planetary systems," *Icarus, 18,* 339-76, 1973.

40. Jaki, S.L. "The five forms of Laplace's cosmogony," *American Journal of Physics, 44,* 4-11, 1976.

41. Jaki, S.L. *Planets and planetarians: A History of theories of the origin of planetary systems*. (Somerset, N.J.: Halsted Press, 1978).

42. Jeans, J.H. "The motion of tidally-distorted masses, with special reference to theories of cosmogony," *Memoirs of the Royal Astronomical Society, 62,* 1-48, 1917(pub. 1923); *Problems of Cosmogony and Stellar Dynamics* (Cambridge, Eng.: Cambridge University Press, 1919); *Astronomy and Cosmogony* (London: Cambridge University Press, 1928).

43. Jeans, J.H. *The Mysterious Universe* London: Cambridge University Press, 1930; rev. ed. 1932, pp. 2-5.

44. Jeans, J.H. "Origin of the solar system," *Nature, 149,* 695, 1942.

45. Jeffreys, Harold. "On certain possible distributions of meteoric bodies in the solar system," *Monthly Notices of the Royal Astronomical Society, 77,* 84-112, 1916; "Theories regarding the origin of the solar system," *Science Progress, 12,* 52-62, 1917.

46. Jeffreys, Harold. *The Earth* (Cambridge, Eng.: Cambridge University Press, 1924), Chapter 1 and Appendixes A & B.

47. Jeffreys, Harold. "Origin of the Solar System," *Nature, 136,* 932-33, 1935; "The origin of the solar system," *Monthly Notices of the Royal Astronomical Society, 108,* 94-103, 1948 (see p. 103); "The origin of the solar system," *Proceedings of the Royal Society of London, A214,* 281-91, 1952 (see p. 290); letter to S.G. Brush, July 3, 1976.

48. Jeffreys, Harold. "Origin of the solar system," *Nature, 153,* 140, 1944.

49. Jeffreys, Harold. "The origin of the solar system," *Monthly Notices of The Royal Astronomical Society, 108,* 94-103, 1948 (see p. 102).

50. Kirchhoff, Gustav Robert. *Untersuchungen über das Sonnenspectrum und die Spectren der chemischen Elemente und weitere ergänzende Arbeiten aus* den Jahren 1859-1862. Osnabrück: Zeller, 1972. See also William McGucken, *Nineteenth-Century Spectroscopy* (Baltimore, Md.: Johns Hopkins Press, 1969), pp. 24-34.

51. Kuhn. W. and A. Rittmann. "Über den Zustand des Erdinnern und seine Entstehung aus einem homogenen Urzustand," *Geologische Rundschau, 32,* 215-56, 1941.

52. Kuiper, G.P. "On the origin of the solar system," in *Astrophysics,* edited by J.A. Hynek (New York: McGraw-Hill, 1951), Chapter 8.

53. Kuiper, G.P. "On the origin of the solar system," *Proceedings of the National Academy of Sciences, 37,* 1-14, 1951; "On the evolution of the protoplanets," *ibid.,* pp. 383-93.

54. Kuiper, G.P. "The formation of the planets," *Journal of the Royal Astronomical Society of Canada, 50,* 57-68, 105-21, 158-76, 1956.

55. Laplace, P.S. de. *Exposition du systeme du monde* (Paris: Imprimerie du Circle Social, 1796). See ref. 40.

56. Lindblad, B. "On the evolution of a rotating system of material particles, with applications to Saturn's Rings, the planetary system and the galaxy," *Monthly Notices of the Royal Astronomical Society, 94,* 231-40, 1934; "A condensation theory of meteoric matter and its cosmological significance," *Nature, 135,* 133-35, 1935.

57. Lyttleton, R.A. "The origin of the solar system," *Monthly Notices of the Royal Astronomical Society, 96,* 559-68, 1936; "On the origin of the planets," *ibid., 98,* 536-43, 1938; "On the origin of the solar system," *ibid., 100,* 546-53, 1940.

58. Lyttleton, R.A. "Dynamical calculations relating to the origin of the solar system," *Monthly Notices of the Royal Astronomical Society, 121,* 551-69, 1960.

59. Miller, S.L. "A production of amino acids under possible primitive earth conditions," *Science, 117,* 528-29, 1953.

60. Moulton, F.R. "An attempt to test the nebular hypothesis by an appeal to the laws of dynamics," *Astrophysical Journal, 11,* 103-30, 1900.

61. Moulton, F.R. "On the evolution of the solar system," *Astrophysical Journal, 22,* 165-81, 1905; *Introduction to Astronomy* (New York: Macmillan, 1906), pp. 463-87.

62. *New York Times.* "Scientific Romancing." October 30, 1949, Sect. IV, p. 8.

63. Parson, A.L. "Vapour pressure of solids at low temperatures (and the origin of the planets)," *Nature, 154,* 707-8, 1944; "The vapour pressures of planetary constituents at low temperatures and their bearing on the question of the origin of the planets," *Monthly Notices of the Royal Astronomical Society, 105,* 244-45, 1945.

64. Payne, C.H. "Astrophysical data bearing on the relative abundance of elements," *Proceedings of the National Academy of Sciences, 11,* 192-98, 1925; *Stellar Atmospheres* (Cambridge, Mass.: Harvard University Press, 1925), pp. 56-57, 185, 188.

65. Plumb, Robert K. "Earth heating up, Urey Theory holds," *New York Times,* October 27, 1949, p. 24.

66. Poincaré, Henri. *Lecons sur les hypotheses cosmogoniques* (Paris: Hermann, 1911; 2d ed. 1913). See also Ref. 8, p. 99, note 233, and ref. 13.

67. Poynting, J.H. "Radiation in the solar system: its effect on temperature and its pressure on small bodies," *Philosophical Transactions of the Royal Society of London, 202,* 525-52, 1903. See also ref. 69.

68. Reuyl, Dirk, and Erik Holmberg. "On the existence of a third component in the system 70 Ophiuchi," *Astrophysical Journal, 97,* 41-45, 1943.

69. Robertson, H.P. "Dynamical effects of radiation in the solar system," *Monthly Notices of the Royal Astronomical Society, 97,* 23-38, 1937.

70. Russell, H.N. "On the composition of the Sun's atmosphere," *Astrophysical Journal, 70,* 11-82, 1929.

71. Russell, H.N. *The Solar System and its Origin.* (New York: Macmillan, 1935).

72. Russell, H.N. "More about the new Lyttleton theory of planetary origin," *Scientific American, 155,* no. 5, 1936, pp. 266-67.

73. Russell, H.N. "Anthropocentrism's Demise," *Scientific American, 169,* No. 1, 1943, pp. 18-19.

74. Saha, M.N. "Ionization in the solar chromosphere," *Philosophical Magazine, series 6, 40,* 472-88, 1920; "Elements in the sun," *ibid.,* pp. 809-24; "On a physical theory of stellar spectra," *Proceedings of the Royal Society of London, A99,* 135-53, 1921.

75. Schmidt, O.J. "A meteoric theory of the origin of the earth and planets," *Comptes Rendus (Doklady) de l'Academie des Sciences de l'URSS, 45,* 229-33, 1944; *A Theory of Earth's Origin, Four Lectures* (Moscow: Foreign Languages Publishing House, 1958). See also L. Randic, "Schmidt's Theory of the origin of visual binary stars and of the solar system," *Observatory, 70,* 217-22, 1950.

76. Slichter, Louis B. "Cooling of the Earth," *Bulletin of the Geolocial Society of America, 52,* 561-600, 1941.

77. Slichter, Louis B. "The Rancho Santa Fe Conference concerning the evolution of the earth," *Proceedings of the National Academy of Sciences, 36,* 511-14, 1950.

78. Spitzer, Lyman, Jr. "The dissipation of planetary filaments," *Astrophysical Journal, 90,* 675-88, 1939.

79. Spitzer, Lyman, Jr. "The formation of cosmic clouds," in *Centennial Symposia, December 1946,* Harvard Observatory Monograph No. 7 (Cambridge, Mass.: Harvard College Observatory, 1948), pp. 87-108.

80. Strand, K. Aa. "61 Cygni as a triple system," *Publications of the Astronomical Society of the Pacific, 55,* 29-32, 1943.

81. Strömgren, Bengt. "The opacity of stellar matter and the hydrogen content of the stars," *Zeitschrift für Astrophysik, 4,* 118-52, 1932; "On the interpretation of the Hertzsprung-Russell Diagram," *ibid., 7,* 222-48, 1933; "On the helium and hydrogen content of the interior of the stars," *Astrophysical Journal, 87,* 520-34, 1938.

82. Strömgren, Bengt. "The growth of our knowledge of the physics of the stars," in *Astrophysics, A Topical Symposium,* edited by J.A. Hynek (New York: McGraw-Hill, 1951), pp. 172-258.

83. Struve, Otto. "The cosmogonical significance of stellar rotation," *Popular Astronomy, 53,* 201-18, 259-76, 1945; *Stellar Evolution* (Princeton, N.J.: Princeton University Press, 1950), pp. 120-53.

84. Sullivan, Walter. "Urey holds moon predated earth," *New York Times,* 27 April 1961, p. 23.

85. Urey, Harold. "A hypothesis regarding the origin of the movements of the earth's crust," *Science, 110,* 445-46, 1949.

86. Urey, Harold. "The origin and development of the earth and other terrestrial planets," *Geochimica et Cosmochimica Acta, 1,* 209-77, 1951; *2,* 263-68, 1952.

87. Urey, Harold. *The Planets: Their Origin and Development* (New Haven, Conn.: Yale University Press, 1952).

88. Urey, Harold. "On the early chemical history of the earth and the origin of life," *Proceedings of the National Academy of Sciences, 38,* 351-63, 1952.

89. Urey, Harold. "On the dissipation of gas and volatilized elements from protoplanets," *Astrophysical Journal, Supplement, 1,* 147-73, 1954.

90. Urey, Harold. "Diamonds, meteorites, and the origin of the solar system," *Astrophysical Journal, 124,* 623-37, 1956.

91. Urey, Harold. "Primary and secondary objects," *Journal of Geophysical Research, 64,* 1721-37, 1959.

92. Van de Kamp, P. "Planetary companions of the stars," *Vistas in Astronomy, 2,* 1040-48, 1956.

93. Weizsäcker, C.F. von. "Über Elementumwandlungen im Innern der Sterne," *Physikalische Zeitschrift, 38,* 176-9, 1937; *39,* 633-46, 1938.

94. Weizsäcker, C.F. von. "Über die Entstehung des Planetensystems," *Zeitschrift für Astrophysik, 22,* 319-55, 1944. For his revisions in the light of subsequent research see the second paper cited in ref. 95, pp. 184-85.

95. Weizsäcker, C.F. von. "Das Spektrum der Turbulenz bei grossen Reynoldschen Zahl," *Zeitschrift für Physik, 124,* 614-27, 1948; "The evolution of galaxies and stars," *Astrophysical Journal, 114,* 165-86, 1951.

96. Whipple, F.L. "Concentrations of the interstellar medium," *Astrophysical Journal, 104,* 1-11, 1946 (presented at Inter-American Congress of Astrophysics, Mexico, 1942).

97. Whipple, F.L. "Kinetics of cosmic clouds," in *Centennial Symposia, December 1946.* Harvard Observatory Monographs, No. 7. (Cambridge, Mass.: Harvard College Observatory, 1948), pp. 109-142. "The Dust Cloud Hypothesis," *Scientific American, 178,* No. 5, 34-44, 1948.

THRESHOLD TO SPACE: EARLY STUDIES OF THE IONOSPHERE

C. STEWART GILLMOR

One of the first applications of space vehicles to science was in the field of upper atmospheric physics, including the ionosphere, those ionized regions above the earth where medium and high frequency radio waves are reflected in their propagation over the earth. Before the era of *in situ* measurements with rockets and satellites, the ionosphere was studied from the ground using radio transmissions. The techniques involved determined a view of the ionosphere which was altered somewhat after the introduction of space research techniques. This paper examines early concepts of the form or profile of electron density above the earth in the ionosphere and the effects of radio measuring instruments upon those concepts. A particular technique may provide only one view.

C. Stewart Gillmor is professor of history at Wesleyan University, Middletown, Connecticut, and Visiting Scholar at the NASA History Office in Washington. He has written on the history of ionospheric research in England, France and the United States.

Recent research in upper atmospheric physics has combined gound-based studies with research using vehicles in space.

The nature of the high upper atmosphere of the earth has been a matter of speculation since well before the 19th century. The aurorae, magnetic variations, and clouds at very high altitudes have given evidence of such an upper atmosphere. With the advent of the technology of radio as a communications medium in the early 20th century, additional evidence was seen for the existence of some kinds of ionized particles high in the earth's atmosphere. While radio engineers worked to improve the quality of radio transmissions and the range and reliability of radio, upper atmospheric physicists (or aeronomists) and some atomic physicists hoped to understand the upper atmosphere: its height, dynamics, and composition.

A third group, using radio as a tool, employed ideas from mathematics and physics to guide them in finding how best to use radio, and similarly, they exploited their radio techniques to discover new aspects of the physics of the earth's atmosphere. In so doing, data produced by their radio tools as well as their earlier *cultural* ideas reinforced a particular idea of the upper atmosphere as being vertically distributed in more or less discrete layers. For reasons of economy and simplicity they sometimes adopted mathematical models which further reified the *layers* which they found in the upper atmosphere.

A partial result of this conceptual modeling was that atomic physicists and aeronomists strove to construct atomic physical models for the phenomenological-cultural artifacts that were layering. The particular tradition of 50 years of radio ideas, the specific technical results of the radio-sounding measuring tools, and the adoption of certain simplified mathematical models led theorists to seek, in the words of one eminent atomic physicist,

> the solution to a problem which can scarcely be said to have existed. We undoubtedly attached too literal an interpretation to the word 'layer'. Perhaps some of us were even misled by the parabolic models used by radio scientists and failed to appreciate fully that they are without real physical significance.[13]

The development of rockets and satellites as tools capable of *in situ* measurements cleared up some of this confusion and ushered in a new era in upper atmosphere and space physics.

Traditional Descriptions of the "Upper Conducting Layer"

I would submit that the atomic physicists and aeronomists were not just *misled* by those who in the 1930s overstated the reality of the ionospheric layers. A concept can owe its existence in large part to the fact that people say it is real. Similarly with the ionospheric layers, from the very beginnings of radio transmission, the ionized upper regions of the earth's atmosphere were compared to real things, mirrors, reflectors, wires, and conducting surfaces.

Following Marconi's demonstration in 1901 of trans-Atlantic radio communication, the radio engineer A.E. Kennelly (born in India of English parents) commented in March 1902 on J.J. Thomson's researches on electrical conductivity of air at low pressure. Kennelly wrote:

> It may be safe to infer, . . . that at an elevation of about 80 kilometers, or 50 miles, a rarefaction exists which, at ordinary temperatures, accompanies a conductivity to low-frequency alternating currents about 20 times as great as that of ocean water. There is well-known evidence that the waves of wireless telegraphy, propagated through the ether and atmosphere over the surface of the ocean, are reflected by that electrically-conducting surface.[34]

In his prescient manner Kennelly observed:

> As soon as long-distance wireless waves come under the sway of accurate measurement, we may hope to find, from the observed attenuations, data for computing the electrical conditions of the upper atmosphere.[34]

Nine months later on 19 December 1902 the great electrical theoretician and misanthrope Oliver Heaviside wrote concerning wireless waves:

> There may possibly be a sufficiently conducting layer in the upper air. If so, the waves will, so to speak, catch on to it more or less. Then the guidance will be by the sea on one side and the upper layer on the other.[30]

While other early workers used various terms,[23] W.H. Eccles suggested the name "Heaviside layer" in 1912. From that time, off and on until the mid-1930s, the ionized layer was called sometimes the "Kennelly-Heaviside Layer" (abbreviated often as K-H-L) and this was taken by most to refer to the entire ionized upper regions, even though Eccles and Arthur Schuster cautioned that the Heaviside layer might only be an edge or portion of the ionized atmosphere.

The pendulum of fashion swings. Until

the mid-1920s some physicists and radio engineers refused even to accept that there was *one* ionospheric layer. By the late 1930s there had been identified or claimed as many as seven different layers! Nevertheless, as Robert Watson Watt said in 1929, "the term 'upper conducting layers' seems to hold the field."[23] I believe this was true in part because of the analogy to optics, in the minds of many, in which the radio waves "bounced" or reflected off the ionosphere as if from a mirror.

Evidence for Layers Produced by Ionospheric Radio Sounders

Ionospheric physics concerns the study of the ionized regions from about 40 km to about 60,000 km above the earth's surface. The larger outer part of this is today called the magnetosphere. Some of the events in ionospheric physics up through the 1950s may be more easily understood if we consider briefly the operation of the ionospheric sounder (now called the ionosonde), the major tool in the study of the ionosphere before the inception of *in situ* measurements from spacecraft. The first experimental radio evidence for an ionized upper layer came from the work of L.F. Fuller and Lee de Forest in 1912-1914. They compared fading between transmissions on early low-frequency circuits and concluded that wave interference was the cause.[44] Others subsequently reported similar findings. T.L. Eckersley and colleagues at the Marconi Company in England made like suggestions from careful study of long distance transmissions on medium-frequency circuits.[22] But these studies did not convince the scientific and engineering world.

The times changed after vacuum tube prices dropped in 1922. Radio became *the* new technical marvel of society. Women's perfume bottles, for example, were designed to resemble vacuum tubes. Radio now was much more than an aid to navigation and a competitor to the telegraph, and broadcasting became big business. The radio

amateur "hams" gained wide publicity with their successes of transmitting very long distances on medium-frequencies with low power. Senior physicists took another look at radio. Joseph Larmor, for example, took up the subject of propagation of radio waves, and his lectures and paper[36] stimulated many to follow, including his junior colleague Edward Appleton, like Larmor a fellow at St. John's College, Cambridge.

The British Broadcasting Company was in its infancy in 1924 but they gladly assisted Appleton in experiments wherein medium-frequency waves of about 385 meter length were received by Appleton about 100 km distant from the transmitter. In these tests, the transmitted frequency was slowly changed. The changing phase differences between the signal arriving along the ground and the signal coming down from the ionosphere allowed Appleton to perform a "Lloyd's mirror interferometer" experiment and estimate the height of the Heaviside layer at about 100 km.[9] This was a great success and the most direct means yet for demonstrating the existence and the height of the K-H-Layer. Others soon repeated these and related experiments and Appleton was off on a long and very productive career which would eventually gain him the Nobel Prize for physics in 1947.

Meanwhile in the United States Merle Tuve and Gregory Breit of the Carnegie Institution in Washington, aided by the Naval Research Laboratory, conducted experiments which were the true beginnings of radar and which were to found the standard technique used in radio sounding of the ionosphere. Breit and Tuve's work,[43] done independently and at about the same time as Appleton's, basically used a transmitter situated very close to a receiver, so that propagation was vertical. The transmitter operated on about 4 MHz (4 Megahertz frequency or 75 meters wavelength) and emitted relatively short pulses of 15 milliseconds length which were sent upward. The receiver then recorded on a string galvanometer the strong direct ground-wave pulse followed by a weaker pulse reflected from the ionosphere. The distance between the two pulse pips on the film record gave a measure of the pulse travel time and thus of the layer height. Tuve and Breit found reflections to occur from 90 to 255 km height. Figure 1 depicts a view of the original equipment used by Tuve and Breit and Figure 2 illustrates one of the earliest records of the pulse method of sounding.

Within eight years the pulse method became the standard in ionospheric studies. Due to a variety of complexities of pulse circuits, multi-stage transmitter and receiver tuning requirements, and antenna bandwidth needs, the early pulse soundings were usually done on one frequency, or they were done alternately on two frequencies. The height of reflection of a radio wave in the ionosphere depends on the concentration of electrons and ions, their mass and charge, and the frequency of the radio wave. Thus for a given condition, a vertically-directed wave of sufficient energy will penetrate upwards until it is bent around and is propagated back to the receiver. Above a certain "critical" frequency, however, the wave sees the ionosphere as transparent and continues upwards into space. Within about two years of his first measurements Appleton, now joined by J.A. Ratcliffe, had discovered a second ionized layer at about 200-250 km.[1] Others found this, too, and at times there seemed to be additional layers whose characteristics were not understood, much less the ion chemistry involved. The properties of the ionization in the ionosphere made it desirable to develop a sounder which could operate continuously over a very wide band of frequencies. If a sounder could be built that would quickly sweep through the wide range of frequencies, then the "critical" frequencies could be measured and the heights of the layers found. This would be a great help in analyzing the possible propagation paths of the radio wave for purposes of prediction.

a

b

Figure 1a, 1b. Original receiving equipment used by Breit and Tuve in 1925 for measurement of virtual reflecting heights of the ionosphere. Fig. 1a shows the receivers, 24 September 1925. Figure 1b shows the power amplifier below and above it the highly damped G.E. Duddell oscillograph (wrapped in white water-cooling hose). Courtesy of Department of Terrestrial Magnetism, Carnegie Institution of Washington.

The first automatic multifrequency ionosphere sounder was developed by T.R. Gilliland at the National Bureau of Standards. Figure 3 shows the very first records from this device, taken 20 April 1933. Several groups in the United States and abroad proceeded to develop variants of the automatic multifrequency sounder. Gilliland's first sounder covered only 2.5 to 4.4 MHz. Later versions covered 1 to 25 MHz. A recent height/frequency record (ionogram) is shown in Figure 4. Note that the "critical" frequency for the F layer appears at a fairly sharply defined boundary, though this need not be the case. It then seems from this that the radio wave is reflected within a narrow height range. A slight frequency change, even a small fraction of a MHz, at the critical frequency of a layer can cause the radio reflection to disappear.

The idea of an ionospheric layer or layers was a useful idea and one subject to elaboration. The physical existence of the layers, however, came to be taken too literally by some. Once the phenomenon has been cast in terms of a metaphor the comparison and resemblance can widen. I will illustrate with some examples taken from associated studies in geophysics. For example, magnetic records that were printed on paper led early workers to see "bays" as if on a geographic map of a coastline. Quick-run magnetic records sometimes contained pulsing phenomena resembling "pearls" on a string. When radio scientists mainly used aural evidence to study very-low-frequency signals they heard "whistlers," "chorus," "tweeks," and "chirps." When later the practice began of reproducing the audio records on paper as visual frequency-time-intensity records, newly discovered phenomena were termed "nose" whistlers, "knees" of curves, and "hooks." Thus the aural metaphor changed to visual. Change in technology of instrumentation and data presentation can cause a change in the way the scientist conceives of the phenomena. Much of the jargon in any technological field is related to man's interactions with nature through his instrumentation.

What I wish to stress is that not only had the term "layer" been in use for decades, now the ionospheric sounder produced a height/frequency plot on which the ionosphere worker could "see" the layers. This

Figure 2a, 2b. Early records made by Breit and Tuve at the Department of Terrestrial Magnetism using the receiving equipment illustrated in Figure 1. The earliest tests used pulses sent at a rate of 500 per second from NKF, a 10 kilowatt transmitter located at the Naval Research Laboratory across the Potomac River at Bellevue, Anacostia, D.C., at a wavelength of 71.3 meters (4.2 MHz). These records were taken 25 September 1925 about 3:30 p.m. EST. The highest humps in the curves are made by waves traveling the shortest path over the ground from the transmitter to the receiver. The smaller humps are caused by reflections from the ionosphere. Figure 2a shows single reflections. Figure 2b shows apparently two reflections following (to the right of) each ground pulse. The distance between the high and low amplitude humps on the curves gives a measure of the reflecting height. Courtesy of Department of Terrestrial Magnetism, Carnegie Institution of Washington.

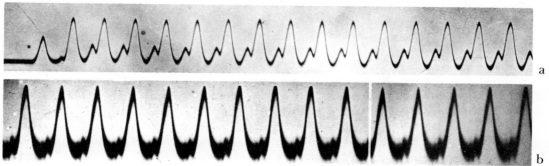

a

b

has certainly been so in my own experiences in ionospheric physics, and I still have this impression as I examine ionograms such as Figure 4.

Models Developed to Fit the Layers

In 1931 Sydney Chapman produced a brilliant attempt[19] to describe how an ionized layer might form in the high atmosphere. He assumed that parallel rays of sunlight of a single color fell upon an upper atmosphere composed of a single type of molecule in isothermal plane-stratified conditions. The sunlight ionized these molecules, producing electrons and positive ions. Chapman assumed that the atmospheric gas became exponentially rarer with height above the earth. Once ionized by the sunlight the electrons remained unattached for some time and then recombined with the positive ions. This compound situation—ionizing radiation penetrating downward and gradually losing energy, and gas density increasing downward—would require a definite maximum of electrons produced at some height. Above and below this height the curve of electron density would fall off. Now, many factors actually intervene in the real terrestrial situation. Ionospheric layers generally arise through quasi-equilibrium between the processes of photoionization, photodetachment, diffusion, mixing and vertical drift.[17] The sun is not visible at night, yet electrons still exist in the ionosphere. The solar spectrum is not monochromatic. The gas composition, temperature and density of the ionosphere were not at all well-known in 1931. Geographic, geomagnetic and solar cycle effects made the picture even more complicated.

Today we can identify numerous ionized regions at different heights which may cause radio reflections from the ionosphere, but we speak mainly of three regions: The D region extends from about 50 to 90 km; the E region from about 90 to 140 km; and the F region from about 120 km up to about 1000 km. These regions may contain one or more

sub-regions or layers depending upon diurnal, seasonal, sunspot-cycle and other factors. See Figure 7 for an indication of the location of these regions as a function of height. Some of the most difficult problems and some of the most clever and imaginative solutions in the 20 years after about 1930 involved theories and experiments to test Chapman's theory of layer formation and maintenance. The main E layer in daytime

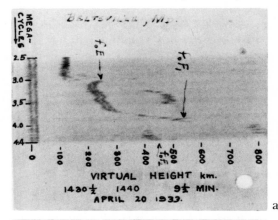

a

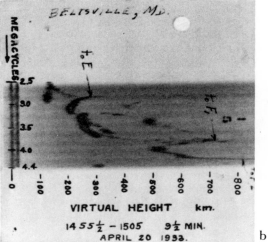

b

Figure 3a, 3b. The first and second ionograms made with the original National Bureau of Standards automatic multifrequency ionosonde at Beltsville, Md., 20 April 1933, over a frequency range of 2.5 to 4.4 MHz. Note the "critical" frequencies, "f_oE," "f_oF_1," & "f_oF_2," indicating reflection from the E and F regions of the ionosphere. Courtesy of the National Bureau of Standards.

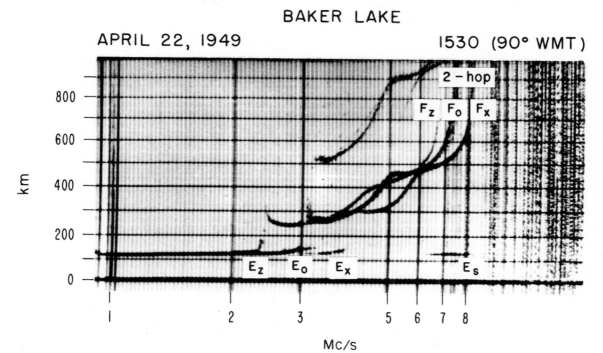

BAKER LAKE

APRIL 22, 1949   1530 (90° WMT)

Figure 4. Sample ionogram (height-versus-frequency record) from Baker Lake, Canada, 22 April 1949. "E_z," "E_o" and "E_x" indicate reflection from the E region. "E_s" is a reflection from "sporadic" sources of ionization such as those associated with aurorae. "F_z," "F_o" and "F_x" are indications of reflections higher in the ionosphere, in the F region. Courtesy of the National Bureau of Standards.

(about 100 km in height) and the F1 layer (about 200 km high) seemed to fit Chapman's theory fairly well. The F2 layer (about 250-300 km high) and at times *all* the other layers, including E and F1, did *not* fit well. An ionospheric sounder receives echoes from layers up to points of maximum electron density. For a unique solution, the electron density must increase monotonically with height; if there are any "valleys" or if the electron density decreases above a maximum there is no echo generally received fron. these regions. The returned echo on any frequency gives information on the integrated electron content over that path. Before the availability of digital computers, such integral equations could be difficult theoretically and nearly impossible to solve in practice. It would seem reasonable, then, to try to fit the data to some simple mathematical curve, and perhaps to adjust the fit with a fudge factor. Thus the ionospheric density profiles have been approximated over the years by linear, quadratic, exponen-

tial, parabolic, and Chapman profiles.[33] See Figure 5. This meant, in fact, that the *lower* side of each layer or region up to the peak of electron density, would be treated as if it were some simple function. This usually works only near the point of maximum density, but for the E layer, and sometimes, in some latitudes for the F1 layer, a parabolic curve can fit both the data as then known and also the ideal Chapman layer. Thus we find Edward Appleton introducing and championing the "Parabolic Layer" model of electron density profile.

Appleton's position regarding the choice of term "region" or "layer" is an interesting one. He is generally credited with first demonstrating not only the existence of the Heaviside or E-layer but also the F-layer (called the "A" or Appleton-layer for a time). Appleton argued for these two layers for a number of years, from 1927[1] until at least 1932, when he wrote his friend Balthazar Van der Pol[5]:

> ... I ought to have mentioned that your suggestion of calling the upper region the A-layer is finding support here, as I think people are now coming to realise that there is a real split between the K-H layer and the upper one. Even the Americans are at last admitting this.

At the same time, Appleton was quite conscious of the danger of over-committing oneself to various narrow layers which might really be a function of season, local conditions, etc. So, beginning in 1929, Appleton carefully used the term "region" whenever possible,[3] for example, as he wrote to J.A. Ratcliffe:[6]

> By the way, in writing up your paper I hope you will use the word 'region' and not 'layer;' but I imagine you already have firm views in that question.

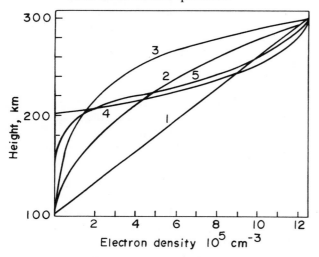

Figure 5. Models of electron density distribution in E and F regions: (1) Linear profile; (2) Quadratic profile; (3) Exponential profile; (4) Parabolic profile; and (5) Chapman profile. (after Kaur *et al.*)

Appleton, like T.L. Eckersley and others, had been much influenced by the great work of Sydney Chapman in proposing the theory of the Chapman layer in 1931. As Appleton wrote to Ratcliffe in 1932,[6] just after publishing his complete magneto-ionic theory paper:

> After getting published the work I have done I don't think I shall be touching magneto-ionics much more. I want to concentrate on the ionisation question and sources of same. ... I have made a list of the possible ways in which U.V. light ... can give us two regions of reflection and would very much like your help in ruling out a few of them.

This interest is seen in several other letters exchanged with Ratcliffe in 1933 and 1934 where they discussed an intermediate region between the E and F regions, and that the intermediate region was "quite strongly ionized." Appleton succeeded in measuring a transition in the reflection coefficient in 1934. This he confided to his friend at Manchester (and former colleague at Cambridge) D.R. Hartree. Hartree replied:[27]

> I am rather surprised, and considerably interested, that you can trace any transition at all; the moral of my paper seemed to me to be that if the thickness of the layer was more than a comparatively small number of vacuum wavelengths, the transition would be pretty sudden. ... Anyhow, the fact that in the reflection coefficient you can observe a transition does suggest that the layer isn't so thick after all. ... I think it would not be difficult to get reflection coefficients for a (parabolic) medium if they would be of interest to you.

So Appleton found evidence that the E layer, for example, could be rather thin and have sharp edges as measured by radio sounding, yet the region itself could pass upwards into the F region through a strongly ionized intermediate space. This could be a controversial position, as he wrote Ratcliffe in 1935:[7]

In confidence, I have received for refereeing a paper by (J.) Hollingworth . . . in which he says that the general theory held is that E and F are quite distinct with no ionization between them. He rejects our work on intermediate region . . .

Appleton's search for appropriate ionization density profiles went on for a number of years. In 1930 he had been assisted, he said,[4] by Jolliffe "our mathematician" for the case of the reflection coefficient $\mu^2 = 1 - A y^n$, where μ = the reflection coefficient, A is a constant, y is altitude, and n is an arbitrary integer. In 1935 he wrote to Ratcliffe:[8]

We ought earlier, I think, to have used this type of ionisation gradient giving $\mu^2 = 1 - \alpha y + \beta y^2$ for this gives us some very interesting results . . . all the other types of gradient we have previously considered don't give a *critical* frequency.

Appleton's concerns continued through 1936. The young Henry Booker was just finishing his Ph.D. and had taught a course titled "Ionosphere" in the Lent Term at Cambridge, in which he treated linear and parabolic layer models.[18] Booker was consulted by Appleton and sent him his work on parabolic regions.[16] D.R. Hartree also sent Appleton his ideas on ionization density profiles, including the parabolic model, and on the transition from the F to the E region, in letters and in a manuscript "Notes on the propagation of electromagnetic waves in a stratified medium." [28,29]

Propagation of the Parabolic Model to the Radio Community

E.V. Appleton gave the Bakerian Lecture to the Royal Society in June 1937 on "Regularities and Irregularities in the Ionosphere."[2] In his lecture he stressed the importance of ionospheric measurements as a valuable means of studying the solar spectrum and the ionosphere as a natural laboratory for atomic physics. He then described the formation of a simple Chapman region, including the variation of ionization density with height. Region F1 during the day, Region E, and Region F2 during winter days, he said, can each closely approximate a simple Chapman region. From radio sounding data Appleton then introduced a model of parabolic distribution with height for ionization density both in Regions E and F. See Figure 6 which reproduces Appleton's 1936 plots of equivalent heights for two parabolic layers. This arises from the assumed ionization distributions shown in the left half of the Figure. It is evident in the Figure that, although Appleton realizes there is considerable ionization in the "valley" between the E and F regions, the ionization supposedly falls to zero on each side of the layers. Whatever his reasons, this Figure published in the Proceedings of the Royal Society, had considerable influence on ionospheric theorists. Compare (Appleton's) Figure 6 with a recent set of profiles measured at sunspot minimum in Figure 7 for the E and F regions during daytime and nighttime.[26]

During World War II, competitive systems were set up within several countries for purposes of predicting radiowave propagations. For example, in Germany and in Britain there were both civil and military prediction services. In the United States the National Bureau of Standards joined the military in what was called the Interservice Radio Propagation Laboratory at NBS in Washington. Newbern Smith of NBS before the War had initiated a prediction system utilizing empirical transmission curves to determine the most efficient frequencies for long-distance communications.[41,42] The British military formed an Inter-Service Ionosphere Bureau (ISIB) headed by T.L. Eckersley and G. Millington of Marconi at Great Baddow, England. This group used an approach similar to the Americans.[38,39] A rival to the ISIB's approach in England was the system developed by Appleton and his younger colleague W.J.G. Beynon at the Radio Research Station at Slough. The most efficient frequencies here were determined from a model employing reflection from a thick parabolic layer. Communications

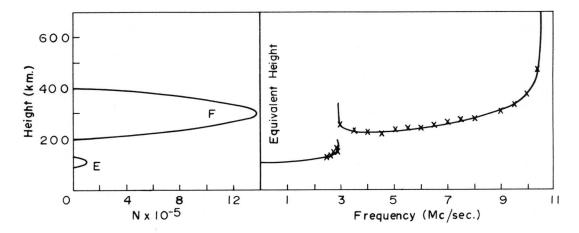

Figure 6. E.V. Appleton's 1937 plots of equivalent heights for parabolic model of two layers, E and F. The assumed ionization distribution is on the left, where "N" represents electron number density (after Appleton).

mixups among the Allies were sometimes laid at the doorsteps of the different prediction models. A partial result of this was that a large and important International Radio Propagation Conference was held in Washington, D.C., in April 1944. Both at this meeting[32] and at a smaller warmup meeting held in London in March 1944[31] the subject of prediction models was a major topic. The Australians were using the parabolic layer method along with Appleton and the RRS at Slough. Meanwhile the ISIB of Eckersley and Millington and the US were using transmission curve methods. Appleton via written report argued that the Appleton-Beynon parabolic method always gave answers within 3% of the American transmission curve method. The Washington conference ended in harmony, yet these conflicts illustrate the importance of the parabolic layer model of electron density profiles of the ionosphere and the firmness with which this model was held.

Assessing the Efforts at Constructing Ionization Profiles

Recently, J.A. Ratcliffe, the foremost liv-ing ionospheric physicist, said of the classical ionosphere:[40]

In the early 1950s, just before experiments were made in space vehicles, there was wide agreement about the distribution of the atmospheric gases up to a height of about 300 km and about the processes by which ionospheric electrons were removed. Experiments in space vehicles confirmed the essential correctness of these ideas.

Ratcliffe admitted that several ideas about layer formation were proved wrong but said:[40]

. . . it is noteworthy that those suggestions . . . that were most widely accepted as reasonable in 1955 were, in fact, shown to be correct when the rocket experiments were made.

Ratcliffe considers the major contribution of rockets and satellites to ionospheric physics to have come mainly in information concerning heights below 90 km and above 300 km.

Yet it seems that others would disagree. D.R. Bates is quoted:[14]

Except for the discovery of radio waves nothing has had such major direct influence on ionospheric research as the advent of sounding rockets and the advent of satellites.

And Bowhill and Schmerling said in 1961:[17]

Although the ionosphere is a very active research field, its study has been to some extent hampered by the fact that the practical applications of radio propagation have led to a phenomenological type of approach to ionospheric behavior. The early availability of large amounts of numerical data on critical frequencies, obtained at the world-wide sounding stations set up for the purposes of long-distance radio propagation prediction, contributed to this approach.

H.S.W. Massey echoes:[37]

The technique of ionospheric sounding used to study the properties of the reflecting layers naturally tended to convey the impression that the maxima were unique and sharply defined. If we look at electron or ion concentration altitude profiles obtained from rocket flights this impression is very much less clear. Because of this the interpretation of factors distinguishing the ionospheric layers sometimes presented a major problem which would

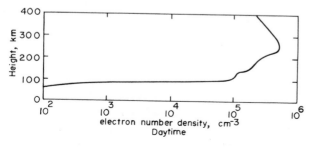

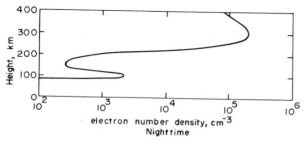

Figure 7a, 7b. Representative electron number density profiles for daytime and nighttime, for comparison with Appleton's parabolic model profile given in Figure 6. Courtesy of F.S. Johnson and Stanford University Press.

perhaps have not appeared so prominent if the basic ionospheric data had come from space flights. Some times in the initial development of a theory it is a disadvantage to know too much and at least the early ionospheric theorists did not suffer in this regard.

Finally, one of the more outspoken younger aeronomists wrote to me in 1972 saying:

. . . the term 'ionospheric physics' is rather meaningless. The charged particles of the upper atmosphere form a very small percentage of the total and are largely at the mercy of the neutral atmospheric components which form the vast majority. It is true there is a cosy, well-knit community of 'ionospheric physicists' . . . but their very narrowness explains their lack of progress in the past 40 years, and it is only recently that the 'fresh air' of general upper atmosphere research has been let in to give their studies a much needed boost.

How can one reconcile these two sorts of opinion, 1) that the classical ionosphere was basically well understood before the era of rockets and satellites; and 2) that the physics of the ionosphere, the composition, formation and dynamics of its constituent particles was little known before *in situ* measurements by rockets and satellites, and that, with respect to layer formation, workers had been misled by overemphasis on discrete layers, especially parabolic models?

Thomas Kuhn is one of those historians of science who believe that "facts" do not stand alone, that a fact has meaning only within the context of a theory or paradigm. As an example Kuhn tells the true story of an investigator who asked a distinguished physicist and an eminent chemist whether a single atom of helium was or was not a molecule. Their answers were not the same. For the chemist the atom of helium was a molecule because of its behavior in the context of the kinetic theory of gases. The physicist, however, saw the helium atom as not being a molecule because it displayed no molecular spectrum.[35]

Thus, each group of upper atmosphere workers may be right. Appleton, Eckersley, Ratcliffe and others at a time of scanty information concerning the earth's upper atmosphere, took the new tools of radio and developed a theory of propagation of radio waves in the ionosphere which explained and allowed for the expansion of medium- and high-frequency radio communications efficiently and reliably over the globe. Some of their theoretical conclusions, while operationally correct, led to incorrect ideas concerning the atomic physics and aeronomy of the upper atmosphere. This was partly due to the technical limits of the sounding instruments developed and utilized importantly for the purposes of long-distance radio propagation predictions. The emphasis on discrete ionospheric layers and the models used to describe these layers did, in fact, prove misleading often to those whose major interest was in studying the formation and composition of the earth's upper atmosphere.

And the era of rockets and satellites, and of high-speed digital computers has not been totally without problems:[17] small probe samples, outgassing from the vehicle walls, ion sheaths produced around the vehicle. Henry Booker remarks[15] that ionospheric wave-propagation studies can be troublesome because with some computer approaches "cost-limitations are very liable to lead to a solution that is presented as 'exact' but is in reality not much better than an appreciably simpler and more analytical treatment that would commonly be described as 'approximate'."

The Future of Ionospheric Physics

In May of 1972 the author administered a mail survey which was responded to by almost 500 persons in over 30 countries. Each person received an eight-page questionnaire containing 27 multi-part questions concerning his career, research interests, professional relationships, etc. Each person in the sample had published at least one paper dealing with ionospheric physics. Collectively, the international respondent group in the decade 1962-1972 decreased its research time in the D, E, and F regions of the ionosphere and increased its research on the magnetosphere, on plasma physics *per se,* and in numerical methods and computation. Each person was asked to write a few sentences describing how the field of ionospheric physics had changed over the time of his career, and what fundamental changes might be anticipated in the next ten years. In the minds of the respondents, outstanding events in the past include the introduction of *in situ* measurements by rockets and satellites, the evolution of computer techniques and electronics, international cooperative projects in geophysics (especially the I.G.Y.), and the development of incoherent backscatter.

Predictions for the ten years 1972-1982 envisioned as probably most important a general unification of theories (and subfields) in atmospheric and terrestrial physics from the upper atmosphere, through the ionosphere, magnetosphere, and solar-terrestrial environment. Both for the geophysicist and for the communications specialist, it was thought that new *techniques* would increase in importance. The major techniques mentioned: rockets, satellites, incoherent backscatter, artificial modification of the ionosphere, use of analog computers for modeling. Many saw the introduction of ground-based techniques in the 1920s as the beginning of the first age of ionospheric physics, and the introduction of *in situ* techniques in the 1950s as the beginning of the second stage.[24]

In recent years the availability of newer types of ground-based instruments combined with changing opportunities for the use of satellites leads one to predict that future successful research on the ionosphere and in other areas of space science will depend upon a combined approach utilizing ground-based methods and laboratory experiments as well as space vehicles.

REFERENCES

1. Appleton, E.V., "The existence of more than one ionized layer in the upper atmosphere," *Nature, 120,* 330-331, 1927.

2. Appleton, E.V., "Regularities and irregularities in the ionosphere," *Proceedings, Royal Society (London), A, 162,* 451-479, 1937.

3. Appleton, E.V., to J.A. Ratcliffe, 18 October 1929. Edinburgh University Library, MS. Gen. 1985/112.

4. Appleton, E.V., to J.A. Ratcliffe, 31 March 1930. Edinburgh University Library, MS. Gen. 1985/120.

5. Appleton, E.V., to B. Van der Pol, 10 May 1932. Edinburgh University Library, Appleton Room.

6. Appleton, E.V. to J.A. Ratcliffe, 24 September 1932. Edinburgh University Library, MS. Gen. 1985/159a.

7. Appleton, E.V., to J.A. Ratcliffe, 23 February 1935. Edinburgh University Library, MS. Gen. 1985/218a.

8. Appleton, E.V., to J.A. Ratcliffe, 24 April 1935. Edinburgh University Library, MS. Gen. 1985/226a.

9. Appleton, E.V. and Barnett, M.A.F., "Local reflection of wireless waves from the upper atmosphere," *Nature, 115,* 333-334, 1925.

10. Appleton, E.V., and W.J.G. Beynon, "The application of ionospheric data to radio communication problems: Part I, "*Proceedings of the Physical Society (London), 52,* 518-533, 1940.

11. Appleton, E.V., and W.J.G. Beynon, "The application of ionospheric data to radio communication problems: Part II, "*Proceedings of the Physical Society (London), 59,* 58-76, 1947.

12. Appleton, E.V., and W.J.G. Beynon, Department of Scientific and Industrial Research. Radio Research, Special Report No. 18, *The Application of Ionospheric Data to Radio Communication,* London, H.M. Stationery Office, 1948.

13. Bates, D.R., "The normal E- and F- layers," *Journal of Atmospheric and Terrestrial Physics, 35,* 1935-1952, 1973.

14. Bates, D.R., "Radiative and collision processes in the ionosphere," *Journal of Atmospheric and Terrestrial Physics, 35,* 2287-2307, 1974.

15. Booker, H.G., "Fifty years of the ionosphere. The early years - Electromagnetic theory," *Journal of Atmospheric and Terrestrial Physics, 36,* 2113-2136, 1974.

16. Booker, H.G., to E.V. Appleton, 4 June 1936. Edinburgh University Library, Appleton Room.

17. Bowhill, S.A., and E.R. Schmerling, "The distribution of electrons in the ionosphere," in L. Marton, ed., *Advances in Electronics and Electron Physics, 15* (Academic Press, New York, 1961) pp. 265-326.

18. Budden, K.G., Personal communication, 24 March 1976.

19. Chapman, S., "The absorption and dissociative or ionizing effect of monochromatic radiation in an atmosphere on a rotating earth," *Proceedings of the Physical Society (London), 43,* 26-45, 484-501, 1931.

20. Chapman, S. and J. Bartels, *Geomagnetism,* 2 vols., Oxford University Press, 1940. Vol. I., p. 504.

21. Davies, K., *Ionospheric Radio Propagation,* NBS Monograph 80 (1965), Washington, D.C.

22. Eckersley, T.L., "The effect of the Heaviside layer on the apparent direction of electromagnetic waves," *Radio Review, 2,* Nos. 2, 5, 1921.

23. Gillmor, C.S., "The history of the term 'ionosphere'," *Nature, 262,* No. 5567, 347-348, 1976.

24. Gillmor, C.S., Unpublished report, sent to all survey respondents, 11 January 1974.

25. Gladden, Sanford C., "A history of vertical-incidence ionosphere sounding at the National Bureau of Standards," National Bureau of Standards, Technical Note No. 28. U.S. Department of Commerce, Washington, D.C., September 1959.

26. Hanson, W.B., *Satellite Environment Handbook,* ed. F.S. Johnson, 2nd edition, (Stanford University Press, 1965), p. 24.

27. Hartree, D.R., to E.V. Appleton, 9 August 1934, Edinburgh University Library, Appleton Room.

28. Hartree, D.R., to E.V. Appleton, 22 February 1936. Edinburgh University Library, Appleton Room.

29. Hartree, D.R., to E.V. Appleton, 6 June 1936. Edinburgh University Library, Appleton Room.

30. Heaviside, O., "Telegraphy," *Encyclopedia Britannica,* 10th ed., *33* (1902) 215.

31. Interservice Radio Propagation Laboratory, unpublished paper IRPL-C2, "Minutes of Discussion on Ionospheric Problems 25th March 1944 at D.S.I.R., Park House, Rutland Gate, London S.W.7." Washington, D.C. April 1944.

32. Interservice Radio Propagation Laboratory, unpublished paper IRPL-C7, "Note on methods of calculating maximum usable frequencies, Sir E.V. Appleton and W.J.G. Beynon, 7 April 1944." Duplicated by IRPL, 20 April 1944.

33. Kaur, K. *et al,* "Phase integral corrections to radio wave absorption and virtual height for model ionospheric layers," *Journal of Atmospheric and Terrestrial Physics, 35,* 1745-1754, 1973.

34. Kennelly, A.E., *Electrical World and Engineer, 39,* 473, 1902.

35. Kuhn, T.S., *The Structure of Scientific Revolutions,* 2nd ed., University of Chicago, 1970, p. 50.

36. Larmor, J., "Why wireless rays can bend round the earth," *Philosophical Magazine,* 48, 1025-1036, 1924.

37. Massey, H.S.W., "Theories of the ionosphere - 1930-1955," *Journal of Atmospheric and Terrestrial Physics, 36,* 2141-2158, 1974.

38. Millington, G., "The relation between ionospheric transmission phenomena at oblique incidence and those at vertical incidence," *Proceedings of the Physical Society (London), 50,* 801-825, 1938.

39. Millington, G., Department of Scientific and Industrial Research and Admiralty, Special Report No. 17, *Fundamental Principles of Ionospheric Transmission,* London, H.M. Stationery Office, 1948.

40. Ratcliffe, J.A., "The formation of the ionosphere, ideas of the early years (1925-1955)," *Geofysiske Publikasjoner, 29,* 13, 1974.

41. Smith, N., "Extension of normal-incidence ionosphere measurements to oblique-incidence radio transmission," *Journal of Research, National Bureau of Standards, 19,* No. 1, 89-94, 1937.

42. Smith, N., "Application of vertical-incidence ionosphere measurements to oblique-incidence radio transmission," *Journal of Research, National Bureau of Standards, 20,* No. 5, 683-705, 1938.

43. Tuve, M.A. and Breit, G., "Note on a radio method of estimating the height of the conducting layer," *Terrestrial Magnetism and Atmospheric Electricity, 30,* 15-16, 1925.

44. Villard, O.G., Jr., "The ionospheric sounder and its place in the history of radio science," *Radio Science, 11,* No. 11, 845-860, 1976.

THE DEVELOPMENT OF AMERICAN LAUNCH VEHICLES SINCE 1945

RICHARD P. HALLION

The twentieth century has witnessed an almost explosive growth in technology, analogous to the rapid advances of the Industrial Revolution. This technology, particularly in the fields of transportation and communication, has led to a point where some futurists now speak of the advanced nations being on the verge of a "post-industrial" society, while others speak hopefully of humanity occupying and colonizing the vast reaches of space itself. In no field has this growth been as rapid and impressive as in the field of aerospace technology. In 1783, humanity first achieved sustained flight using balloons. But it was not until 1903 that the first successful airplane appeared, and not, in fact, until the late 1920s that air transportation of goods and passengers became a practical reality. Likewise, following the medieval introduction of the blackpowder rocket, rocket technology stagnated with

Richard P. Hallion is associate professor at University College, University of Maryland, and author of several books in aerospace history.

but modest advances until, in 1926, Robert Goddard successfully demonstrated a liquid-fuel rocket. From Goddard onwards, advances by rocket technologists around the world made possible practical manned and unmanned spaceflight, including (less than five decades after the beginnings of modern rocketry) the first manned flights to a celestial body.

A persistent theme in the history of technology is the apparent lack of foresight among established scientists and technologists with regard to technological revolutions that are brewing, as it were, just around the corner. The steam-driven locomotive and ship, the airplane, the jet engine, to name just a few, were all regarded skeptically and even decried as technologically impossible and socially and economically undesirable until after each had been actually demonstrated. And so it was with the rocket. In 1939, when he was offered an opportunity to enter the field of rocket research or leave it to the Guggenheim laboratory at the California Institute of Technology, Jerome C. Hunsaker, director of aeronautics at the

Massachusetts Institute of Technology told Caltech's Theodore von Kármán, "You can have the Buck Rogers job."[33] Vannevar Bush, later the director of the National Advisory Committee for Aeronautics, once alledgedly told von Kármán and Caltech president Robert Millikan "I don't understand how a serious scientist or engineer can play around with rockets."[34] Yet today those who played around with the Buck Rogers job have created a technology that has enabled the exploration of space, and that fills the arsenals of major nations. It is a technology that can enrich our existence, or, if misused, bring that existence to an end.

Even before the clamor of Sputnik had died away, the utilization of orbiting platforms for military and scientific purposes had begun. But placing satellites into orbit, and, later, sending space probes to the moon and distant planets, demanded the development of reliable booster launch systems. As with the earlier history of winged aviation, one can trace a "leader-follower" relationship between the development of military rocketry and the application and adaption of this technology for non-military purposes. The first space boosters were all spin-offs from existing military missile systems, and, indeed, the roots of this technology, almost without exception, extended to German wartime research at Peenemünde.

Using rockets to boost scientific payloads into space first appeared in the conceptual writings of Konstantin Tsiolkovskii, Robert Goddard, and Hermann Oberth, whom rocketeer G. Edward Pendray has credited with being "The three great progenitors of the modern space age."[22] Later, a host of lesser-known (but in some ways more influential) technologists such as Max Valier, Walter Hohmann, Eugen Sänger, Mikhail Tikhonravov, Sergei Korolev, James Wyld, Wernher von Braun, and Willy Ley advocated rocket-boosted scientific experiments in space. Robert Goddard was well on the road to developing powerful liquid-fuel rockets, but allowed his research to become

sidetracked in peripheral issues and narrow specialties, so that it did not reach practical fruition until others extended it after his death in 1945. The American Rocket Society (previously called the American Interplanetary Society), the German *Verein für Raumschiffahrt* (Society for Space Travel), and the Soviet *GIRD (Grupi Izucheniyu Reaktivnogo Dvizheniya*—Group for the Study of Reactive Motion) all undertook ambitious research during the 1930s for the purpose of developing rocket technology that would permit the firing of payloads into space.[35]

The V-2 and Postwar American Rocketry

With the development of the German V-2 ballistic missile, researchers had at last a booster system capable of launching a meaningful scientific payload into the upper atmosphere. Overtly developed as a military weapon, the V-2 had a range of approximately 200 miles, and was fired in a terror campaign against Allied population centers in 1944-45, causing numerous civilian casualties. Following the collapse of Nazi Germany, Allied technical investigators acquired both production examples of the V-2 and its equipment as well as members of the German rocket development teams.[9] Many of these individuals contributed to the success of subsequent American, Russian, British, and French space programs, the largest number coming to the United States under the aegis of Project Paperclip.[17]

The V-2 effort served as the basis of the first major postwar development programs in both the United States and the Soviet Union.[21] In 1946, the United States launched the first of 67 captured V-2's to be fired in America, during trials at White Sands, New Mexico. Over a year later, the first Soviet V-2 roared away from its launch site at Kapustin Yar. Modified V-2's and V-2 "spin-offs" served as the basis of both American and Soviet upper-atmosphere sounding rockets, as well as inspiring the first generation of American and Soviet medium-range ballistic missiles.[6,7] The great potential of the rocket as a research tool led to the Federal government's creation of an interagency Upper Atmosphere Rocket Research Panel under the direction of James Van Allen.[25] However, the United States was not totally dependent upon the V-2. Major indigenous programs existed, inspired both by the realization that the supply of V-2's would not suffice for more than a few years, and by advocates who championed purely American efforts. American postwar rocket research drew upon traditional governmental-industrial partnerships, as well as upon such newly established bodies as the Jet Propulsion Laboratory (JPL) and the RAND corporation. Figure 1 provides a listing of selected postwar American rocketry programs from 1946 through 1958, showing the principal agency and the purpose of each program. It must be mentioned that a dense bureaucratic web surrounded most of these projects, and that administrative responsibility and technical mangement were often shared between several organizations. For example, the highly successful Aerobee was inspired by the JPL, sponsored by the Navy's Bureau of Ordnance, designed and constructed by Aerojet, assembled for service by Douglas, and supervised technically by the Applied Physics Laboratory of Johns Hopkins University![20] The entangled procurement networks of today's space program extend back far beyond the carefree days of the 1960s.

Many of the military weapons systems listed in Figure 1 furnished the basis for subsequent space and sounding rocket launch vehicles, such as Redstone, Nike, Atlas, Thor, and Titan. With the exception of solid-fuel propellant systems favored for most small rockets, American and Soviet rocket developers expressed a marked preference for using essentially the same propellants that had been favored by pre-war and wartime researchers: liquid oxygen and diluted alcohol or a petroleum-base fuel. Others used such already-traditional combinations as red fuming nitric acid and aniline,

and some held out for such future propellants as unsymmetrical dimethylhydrazine (UDMH) or liquid hydrogen. Generally, however, the Peenemünde tradition of "lox" and alcohol, or, say, kerosene, held firm for the early years. As years went by, however, the desirability of using such exotic fuels as liquid oxygen for operational military missiles appeared increasingly questionable, especially since it involved fueling the missile immediately prior to launch, increasing its vulnerability and reducing potentially critical reaction time. For these reasons, military planners increasingly favored long-term storable liquid propellants, and, eventually, solid-fuel propulsion systems as incorporated on the American Polaris, Poseidon, and Minuteman missiles, and such Soviet missiles as the SS-13, SS-14, and SS-20. But such requirements meant little for boosters involved in the exploration of space, with the result that, today, derivatives of such earlier liquid-fuel missiles as the Atlas and Thor still enjoy wide usage and popularity for launch missions.

Numerous technical problems confronted missile developers in the 1940s and 1950s, some of which required immediate solutions for military projects, but which were not so urgent for scientific launch programs. Development of reliable guidance systems, especially inertial guidance, proved especially difficult, costly, and time-consuming. Turbopump technology (to furnish rocket engines with propellants) involved a great amount of research into materials and propulsion schemes for the pumps. Combustion instability plagued early large rocket engines, often resulting in destructive explosions that destroyed missiles and engines in spectacular eruptions that hampered subsequent accident investigations. Reentry proved a difficult area, even following the postulation of the blunt-body reentry principle by NACA scientist H. Julian Allen in the early 1950s, for different opinions existed on the types and kinds of thermal protection systems that might be used to pro-

SELECTED AMERICAN ROCKETRY PROGRAMS, 1946-1958

Agency	Programs	Purpose
JPL	Private A/F	RRV
	WAC Corporal/B	SR
	Sergeant (small)	RRV
	Corporal	SSM
	Sergeant	SSM
	Loki	SAM, SR
Army	Redstone	SSM
	Jupiter A	RRV
	Jupiter C	ReR
	Juno I	SLV
	Jupiter	IRBM
	Juno II	SLV
	Honest John	SSM
	Nike-Ajax	SAM
	Nike-Hercules	SAM
Navy	Viking	RRV, SR
	Vanguard	SLV
	Aerobee	SR
	Arcon	SR
	Polaris	IRBM
Air Force	Atlas	ICBM
	Titan	ICBM
	Thor	IRBM
	MX-774	RRV
	Minuteman	ICBM
	X-17	ReR
NACA (Nat'l Advisory Committee for Aeronautics)	Deacon	SR
	Nike-Deacon, Nike Cajun	SR
	Honest John-Nike	SR
Univ. of Maryland	Oriole	SR
	Terrapin	SR
Univ. of Michigan	Exos	SR

Key: RRV—rocket research vehicle.
     SR—sounding rocket
     SSM—surface-to-surface missile.
     SAM—surface-to-air missile.
     ReR—reentry research.
     SLV—satellite/space probe launch vehicle.
     ICBM—intercontinental ballistic missile.
     IRBM—intermediate range ballistic missile.

Figure 1

tect a nose cone during atmospheric entry. (Eventually this technology, developed initially to protect nuclear warheads while they transited the atmosphere on the way to their targets, would be utilized to protect astronauts and cosmonauts returning to earth, as well as spacecraft entering other planetary atmospheres). Yet with persistence these difficulties were overcome, so that, by October 1957, the Soviet Union felt confident enough to launch Sputnik, and by the end of 1958, an entire Atlas missile had been boosted into orbit as Project Score. Figure 2 gives a listing of selected missiles and boosters with the thrust rating of their propulsion system.

An important step occurred on February 24, 1949, with the launch of the so-called Bumper WAC from White Sands. Bumper WAC utilized a JPL-Army Signal Corps WAC Corporal sounding rocket as the second stage of a V-2, an idea first advocated by Martin Summerfield and Frank Malina of the JPL.[18] Bumper WAC reached an altitude of 244 miles, the first rocket to enter extraterrestrial space, and the "ancestor" of all subsequent multi-stage American rockets.[19]

Launch Vehicle Development: The Military Perspective

The three major military services all sponsored separate missile and rocket development efforts that eventually proved crucially important to the evolution of American launch vehicles. The Army, the service most closely influenced by the former Peenemünde team, sponsored development of the Restone missile which eventually spawned the Jupiter C and Juno I, Jupiter IRBM and Juno II, and, via the Juno III, IV, and V paper studies, the Saturn launch vehicle

PROPULSION CHARACTERISTICS OF SELECTED MISSILES AND BOOSTERS

Vehicle	Year	Fuel*	Stages	Thrust (lbs.)*
V-2	1944	Liquid	One	60,000
Private A	1944	Solid	One	1,000
WAC Corporal	1945	Liquid	One	1,500
Corporal	1947	Liquid	One	20,000
Viking	1949	Liquid	One	20,450
Redstone	1953	Liquid	One	75,000
Jupiter C	1956	Liquid	Three	83,000
Thor	1957	Liquid	One	150,000
Jupiter	1957	Liquid	One	150,000
Vanguard	1957	Liquid	Three	27,000
Juno I	1958	Liquid	Four	83,000
Atlas B	1958	Liquid	Stage-and-a-half#	360,000
Juno II	1958	Liquid	Four	150,000
Tital I	1959	Liquid	Two	300,000

*For first stage only. Some had solid-fuel upper stages, and upper stage thrust figures were, of course, also different.

#i.e.: combined twin booster engines with a sustainer engine for lift-off.

Figure 2

family that placed astronauts on the moon before 1970. The Navy pursued development of the Viking sounding rocket, furnishing much of the technology that later appeared in the Vanguard satellite launch system. (Vanguard-derivative upper stages later appeared on the Scout, Able, and Delta boosters). Additionally, the Navy's Polaris solid-fuel submarine-launched ballistic missile program generated components incorporated in the NASA Scout launch vehicle. The Air Force pursued development of long-range intercontinental ballistic missiles starting in 1946 with research project MX-774, then project Atlas, which eventually led to the evolution of the Titan and Minuteman ICBM family, as well as the Thor IRBM companion effort. With the exception of Minuteman, Atlas, Titan, and Thor derivatives have all appeared as major American standard launch vehicles.

The Army's Redstone program could be considered a "second generation" V-2, for there were marked technological similarities between the two missiles, and the development team was largely the same. Redstone, a medium-range (200 mile) ballistic missile, burned a mixture of liquid oxygen and diluted alcohol in a single-chamber rocket engine producing 75,000 lbs. thrust for 121 seconds. It stood 69 feet in height, and weighed approximately 61,000 lbs. at launch. The Army-von Braun team initiated development of the Redstone in 1950 (as part of the earlier Hermes C-1 effort), and it first flew in 1953 at Cape Canaveral. Redstone demonstrated the first American inertial guidance system, and led to the Jupiter C, Juno I, and Mercury-Redstone launch vehicles. Jupiter C had first been proposed by the von Braun team in 1954 for the anticipated "Project Orbiter" satellite effort, but when, in 1955, the Department of Defense threw its full support behind the Naval Research Laboratory's Project Vanguard, the Army shelved Project Orbiter, and, instead, used the Jupiter C for reentry nosecone tests in support of the Jupiter

IRBM effort (the Jupiter IRBM was a much larger missile). The three-stage Jupiter C (with JPL-developed upper stages) first flew in 1956. Following the orbiting of Sputnik I by the Soviet Union in 1957, the Department of Defense reversed its earlier decision, and authorized the Army Ballistic Missile Agency (later largely absorbed into the new National Aeronautics and Space Administration) to proceed with the launching of a small earth satellite during the International Geophysical Year. On January 31, 1958, a Jupiter C derivative, the Juno I, launched Explorer 1, the first American satellite. Juno I differed from the three-stage Jupiter C only in having a small solid-rocket motor as a fourth stage. The Redstone still had one major task to perform in America's space program: launching the first suborbital tests of the manned Mercury spacecraft. NASA lengthened the missile to provide greater propellant volume and to accomodate the McDonnell spacecraft and escape tower. Following two check flights (the second of which carried the chimpanzee Ham into space and returned him safely to earth), astronaut Alan Shepard completed a test flight to 115 miles altitude and 302 miles downrange on May 5, 1961. Over two months later, Virgil "Gus" Grissom followed Shepard into space, also on a Redstone-derived booster.[32]

The Army's Jupiter IRBM program rivalled the Air Force's Thor program, and even involved the Navy, for the missile was to be both land and sea-based. Eventually, the Navy dropped out in favor of the solid-fuel Polaris—the Jupiter, like Redstone, was a liquid-fuel missile—and the Jupiter became embroiled in an internecine struggle with the Air Force's Thor for official support. Though both won production contracts, Thor achieved greater prominence as a launch system.[1] The single-stage Jupiter did spawn the Juno II launch vehicle, with a Jupiter first stage, and Juno I upper stages, which enjoyed an approximately 45% success rate for attaining mission goals during ten launches in the 1958-1961 time period.

Figure 3. Bumper—WAC at Cape Canaveral c. 1952.

The Juno II did launch the Pioneer 3 and 4 deep space probes and the Explorer 7, 8, and 9 "heavy" earth satellites. Five of the ten launch attempts were total failures. The Juno II program led to the Juno III, IV, and V studies preceding the Saturn program, which is discussed subsequently.

The Navy's postwar rocketry programs centered on the Viking sounding rocket, satellite launcher studies, and some form of sea-launched ballistic missile. Similarly to their Air Force counterparts, Navy rocket advocates often found themselves locked in competition for support and resources with those who favored long-range jet-propelled winged cruise missiles such as the Regulus I and II (and the Air Force Matador, Snark, and Navaho), much as wartime German V-2 supporters had clashed with those in favor of the jet-propelled V-1. The Viking program owed nothing to the Peenemünde experience, and pioneered, for example, a gimbaled engine for steering, an aluminum structure, and post-boost gas jet operation to maintain a desired attitude during the coast to maximum altitude (as high as 158 miles). Viking, a joint Naval Research Laboratory (NRL)-Martin effort, complimented an earlier study by the Navy's Bureau of Aeronautics in 1946 for a High Altitude Test Vehicle (HATV) capable of placing a 1,000 lb. payload in earth orbit. Eventually the Navy launched a series of 12 Viking rockets of various configurations, and these single-stage craft furnished quite useful atmospheric and space sciences information for the time.[26,27]

More importantly, the Viking led directly to the three-stage Vanguard, the officially sanctioned satellite launch system that performed so disappointingly when first called upon to attempt an orbital launch.[11] Essentially, the Viking formed the basis for the Vanguard's first stage, while the Aerobee sounding rocket inspired the second stage, and the third stage consisted of a new specially designed solid-fuel rocket engine.[14] (This example of building up a launch system from largely existing hardware has been traditional in most of the world's major space booster efforts). Vanguard was still undergoing development when the post-Sputnik furor caused an acceleration of launch plans. In response to a poorly timed White House announcement, program officials had to ready the first complete Vanguard, TV-3, for a hasty launch before a public largely believing that Vanguard was a thoroughly proven system. On December 6, 1957, Vanguard lifted off, fell back, and exploded on its launch pad, tarnishing the program with the image of total failure. Vanguard 1 followed the earlier Explorer 1 satellite into orbit on March 17, 1958. Two other Vanguard satellites followed. Its greatest contribution, however, was in Vanguard derivatives for other programs, including Able, Scout, and Delta, all of which incorporated

Vanguard upper stages.[28]

It is the Air Force programs, however, that most often come to mind when mention is made of the major launch vehicle programs of the 1950s and early 1960s: Atlas, Thor, Titan, and Minuteman. In the immediate postwar years, a key group of Air Force planners worked to develop intercontinental nuclear-tipped missiles. They faced criticism from opponents who favored continued reliance upon manned bombers, supporters of long-range pilotless jet bombers similar in general concept to today's cruise missile, and from critics (such as Vannevar Bush) who questioned the technological feasibility of intercontinental rocket bombardment systems. Faced with budgetary restraints, official reluctance, and the technological problems inherent in developing any new technology, rocket advocates moved most cautiously.[3] In a manner analogous to the actions of strategic bomber supporters in the 1930s, they proceeded slowly, emphasizing technological problem-solving and marshalling arguments in their favor, notably the vulnerability of manned bombers and unmanned pilotless jet aircraft to defensive countermeasures. Then, in the early 1950s, a string of political and scientific-technological events occurred to place ballistic missile development in a more favorable official light: perceptions of a threat from the Soviet Union, the postulation of the blunt-body reentry concept, the development of small high-yield thermonuclear weapons, and the fortuitous placement of two key missile advocates within the government's highest echelons of defense planning, John von Neumann of Princeton's Institute for Advanced Study, and Trevor Gardner, special assistant to the Secretary of the Air Force.[11] As a result, in 1954 the Air Force issued two "General Operational Requirements," one for the Atlas ICBM, and the other for the Thor IRBM.[8]

By the end of 1955, the Air Force had no less than four demanding missile and launch programs underway: the intercontinental Atlas ICBM (WS 107A-1); Titan (WS 107A-2), considered a less-risky backup for Atlas; the intermediate-range Thor IRBM (WS 315A); and, finally, an Atlas-boosted upper stage satellite launch system that eventually evolved as the Agena, for launching a reconnaissance satellite system designated WS 117L. These missile programs ran concurrently with two major cruise missile efforts, Navaho and Snark; the Air Force expected that Atlas would not be ready for operational deployment before 1965, and that the national nuclear deterrent would consist of these cruise missiles until Atlas became operational. In fact, there was a great deal of technical interchange between the programs whenever possible; Atlas and Thor, for example, used a main engine developed originally for the Navaho booster that would accelerate the ramjet-powered Navaho cruise missile to supersonic speeds. Management of this development effort was of critical importance. As program analyst Robert Perry has written:[23]

> the *management of technology* became the pacing element in the Air Force ballistic missile program. Moreover—as had not been true of any earlier missile program—technology involved not merely the creation of a single high-performance engine and related components in an airframe, but the development of a family of compatible engines, guidance subsystems, test and launch site facilities, airframes, and a multitude of associated devices.

Atlas was officially regarded as a high development risk program, especially in the fields of propulsion, guidance, and design of its reentry warhead. To reduce weight to a minimum, Convair engineers under the leadership of Karel Bossart designed the craft with a very thin internally pressurized airframe (like a balloon) having integral propellant tanks. As time went on, however, Atlas appeared less radical and risky, and the

Figure 4. Failure of Vanguard III, December 6, 1957.

backup Titan accordingly grew more sophisticated.[24] Titan, with its monocoque airframe (like a modern airplane fuselage), was a two-stage missile as opposed to the three-engine "stage and a half" Atlas. Further, it featured a more advanced guidance system. Eventually (on the Titan II), Titan incorporated storable hypergolic (self-igniting) propellants and a completely inertial guidance system, making it especially well-suited as a strategic rapid-response missile system, in contrast to the (by then conservative) liquid-fueled Atlas—and earlier Titan I—which required fueling prior to launch.

Thor, a deliberately rushed program, nevertheless proved remarkably successful, taking only 3.5 years from program approval to delivery to its first operational squadron. Indeed, Atlas, Thor, and Titan

Figure 6. Atlas-Centaur

Figure 5. Thor

Figure 7. Saturn V

proved remarkably rapid programs, in contrast to such earlier manned bomber efforts as the B-47 and B-52, and especially in contrast to the 1950s generation of intercontinental cruise missiles, Snark and Navaho. Snark eventually attained only limited service, and Navaho was abandoned. The single-stage Thor completed its first "full-range" test flight on September 20, 1957. The Atlas reached its own full-range milestone over a year later, on November 28, 1958. Both Atlas and Thor went on to highly successful careers as "workhorses" of the space age, launching a variety of payloads, using appropriate upper stages, such as the Centaur and Delta. The Atlas D served as the basis for the Mercury-Atlas launch vehicle which boosted the first Mercury astronauts into orbit, beginning with John Glenn on February 20, 1962.[31] The Titan II lofted the Gemini spacecraft into orbit later in the decade.[13] Eventually, the Titan program spawned the Titan III family of heavy launch vehicles. The only major Air Force missile system not resulting in a suitable space launch vehicle derivative was Minuteman, a much smaller solid-fuel ICBM that has since become the principal American strategic missile system (as of 1980, the United States Strategic Air Command deploys 1,000 Minuteman and Titan II ICBM's in launch silos across the Midwest and Western United States). Government-sponsored studies have indicated that Minuteman could easily be modified for a launcher role, especially once it is superseded by the planned MX missile in its military mission.[16] Until such modifications are authorized, however, Minuteman remains the exception to the rule that American military hardware has invariably spunoff space boosters out of overtly destructive missile systems.

Scout and Saturn: Two NASA Stories

Two programs offering interesting comparisons and contrasts that were civilian efforts were the Scout and the Saturn. The former, a "lash-up" of largely existing components, has evolved as a highly reliable launch vehicle for small scientific and research payloads. The later was, of course, the largest (to date) liquid-fueled rocket ever developed. Scout involved basically one NASA center and field site for its development, while Saturn involved many centers and contractors, thousands of individuals, and billions of dollars.

Scout began in 1957 as a study by the Pilotless Aircraft Research Division of the NACA's Langley laboratory, later the NASA Langley Research Center, to boost the speed of solid-fuel rockets to orbital velocities. Researchers investigated various solid-rocket systems, but finally decided to combine an Aerojet Jupiter Senior (a 100,000 lb. thrust engine burning ammonium perchlorate-polyurethane and aluminum) together with an improved Thiokol Sergeant for the second stage, and two new upper stages descended from the Vanguard. The eventual result was a shapely missile dubbed "Scouts" (and later just "Scout"), the acronym standing for Solid Controlled Orbital Utility Test System, with its four stages poetically named Algol (first), Castor (second), Antares (third), and Altair (fourth). Despite official reluctance initially, NASA eventually authorized a full go-ahead on the launch system in mid-1958. It could place a 155 lb payload into a 340 mi. orbit, or 330 lb payload into a 115 mi. orbit. NASA fired its first complete Scout on July 1, 1960, and despite some initial failures, the missile proved quite successful, achieving an overall mission successful launch rating of 85% by the end of 1968. Many uses have been found for this adaptable vehicle; indeed, during the mid-1960s, some researchers even examined the possibility of further boosting the Scout's performance by firing it from a climbing X-15 rocket research airplane (and the X-15 itself was, of course, air-launched from a modified B-52). Scout will always be judged as an excellent example of the successful application of off-the-shelf technology to meet a specific need—in this case a cheap, flexible, and reliable launch system for small payloads.[29]

Saturn, on the other hand, will always be synonymous with massive application of funding and technology to achieve a gargantuan task: developing a large and complex liquid-fuel launch vehicle capable of placing a team of astronauts on the moon and then returning them safely to earth. The Saturn vehicle posed numerous design challenges requiring creative and insightful engineering analysis, especially in the fields of propulsion technology, overall system reliability, structures, and cryogenics. The origins of the Saturn program actually predated Sputnik—and, indeed, a claim can be made that they extend all the way back to the large boosters fancifully envisioned by von Braun's engineers at Peenemünde when they were not busily working on the V-2. In 1957, the von Braun team at the Army Ballistic Missile Agency began studies of a 1.5 million lb. thrust launch vehicle. As with other American and Soviet booster developers, the study team quickly opted for a clustered approach using smaller engines, and this received the name Saturn I. Plans to use off-the-shelf engines modified to provide higher thrust soon crumbled under the impact of tremendous technical difficulties, most of which, as historian Roger Bilstein has related, were solved by "cut and try" methods.[2] In 1959, a joint NASA-Department of Defense Saturn evaluation committee opted for using liquid hydrogen as a high energy upper-stage propellant. Four years later, the Saturn program had standardized on three major launch vehicles, the I, IB, and V. Only the latter was capable of boosting a lunar mission. The others were strictly orbital boosters. All used liquid hydrogen as upper stage propellants; fortunately for NASA, the agency could draw on experience with Convair's Centaur, a pioneering liquid hydrogen upper stage. Weight control became a critical concern. The first-stage Rocketdyne F-1 engine, de-signed to produce 1.5 million lbs thrust, proved particularly troublesome, causing one engineer to characterize rocket engine technology as a "black art." Finally, following intensive testing and modifications, the F-1 proved reliable enough so that five could be clustered together in the first stage, furnishing the Saturn V with an awesome 7.5 million lbs. of thrust at lift-off. In 17 unmanned and 15 manned launches, the Saturn booster family scored an impressive 100% launch success rate, to the relief of program observers who were well aware of the many technical challenges that had stood in the pathway to success. Saturn's shining moment, of course, would always be July 16, 1969, when it launched Apollo 11 on its historic journey to the moon.

Saturn demonstrated that with tight program control, a solid technical base (building, literally, on the old V-2 and Redstone experience, up through Jupiter C, Jupiter, and the Juno family), tremendous funding and material support, and sufficient manpower, that a complex system could be made to function reliably on the first try. But it also demonstrated the massive and almost impractical size and power requirements of chemically fueled earth-launched rocket systems for manned spaceflight, even for such a goal as a trip to our nearest celestial companion.

Mix 'n Match: Upper Stages + Reliable Lowers = New Opportunities

The lack of heavy-lift boosters in the early American space program, more than any other single factor, limited our early exploration of the earth and solar system from space, and contributed to the myth of a "missile gap" between the United States and the Soviet Union.[4] One solution was the addition of powerful upper stages to existing missiles such as the Thor and Atlas. The first major upper stage development effort was the Air Force's Agena program. Agena, manufactured by Lockheed, began as a boost system for an advanced military satellite. In January

1959, the newly formed NASA announced the agency's intention to use Agena as an upper stage with Atlas and Thor missiles. Agena used a Bell-developed rocket engine burning a mix of inhibited red fuming nitric acid and unsymmetrical dimethylhydrazine (IRFNA and UDMH). It could boost a 2,200 lb. payload to a 115 mi. earth orbit. Agena development was rapid enough so that on February 28, 1959, the Air Force was able to use a Thor-Agena combination to launch Discoverer 1, the first Air Force satellite. A follow-on, the Agena B, flew on both Thor and Atlas first stages, boosting such satellites and payloads as Nimbus 1 (Thor-Agena), and the Ranger lunar spacecraft (Atlas-Agena). Agena B eventually gave way to Agena D, still used with modified Titan, Thor, and Atlas first stages.[10]

Centaur, another very successful upper stage program, grew out of interest in using liquid hydrogen propellants for rockets—not a new idea, but one that posed considerable challenges. Under the leadership of Krafft Ehricke, a Convair engineer, the company began studying a high-energy upper stage that could be flown on the Atlas. With support from the Advanced Research Projects Agency, the Air Force, and Pratt & Whitney (manufacturers of the Centaur's engine), Centaur moved rapidly ahead. It benefited from a previous research project dubbed Suntan for a long-range supersonic hydrogen-fueled reconnaissance airplane and from studies on liquid hydrogen undertaken by NASA's Lewis research facility. Despite a long development program, Centaur completed a successful test flight on November 27, 1963, by which time, responsibility for its development had been transferred from the Air Force to NASA. By 1966, following further test flights, Centaur was ready for operational use, and over the next several years it launched such spacecraft as the Surveyor lunar probe series and OAO 2. The development experience with Centaur

Figure 8. Delta 3914.

helped NASA anticipate many of the problems later encountered during the development of the liquid hydrogen upper stages for the Saturn booster family. Centaur has gone on to become one of the major American upper stages used with modified Atlas and Titan III first stages.[30]

The greatest "success story," however, involves the Thor-Delta, frequently referred to as simply the Delta. NASA desired to modify the Thor so that it could serve as a medium-payload boost system until larger capacity launch vehicles became available. To do this, the agency foresaw using proven Vanguard hardware (specifically, the Vanguard X-248 third stage mounted on a modified Vanguard second stage). This strap-together combination was ready for its first launch by mid-1960, and on August 13, 1960, the first successful Thor-Delta boosted the Echo 1 passive communications satellite into orbit. When the proposed Vega upper stage program of NASA was cancelled, the agency decided to continue using the Thor-Delta combination. Now, however, the Thor-Delta underwent an extensive growth period, becoming the single most versatile and significant booster of the 1960-1980 time period. A partial listing of payloads lofted into space by Thor-Delta vehicles reads like a "Who's Who" of the early space age: Echo, Tiros, OSO, Ariel, Telstar, Relay, Syncom, Intelsat, ESSA, Biosatellite, Landsat, Symphonie, GOES. Thor-Delta combinations, by the time of the series' 150th launch (in 1979) had achieved an impressive average of eight launches per year with a success rate of 92%. The vehicle benefitted from the addition of solid-fuel "strap-on" rockets around the Thor first stage (sometimes as many as nine solids of the Castor variety), as well as improved upper stages from the Scout, and increasing the length of the basic Thor vehicle itself. Thor-Delta went from an ability to place 600 lbs into a 155-mile high orbit in 1960 to 4,300 lbs in

Figure 9. Titan IIIE-Centaur

such an orbit by 1972.[16]

Currently, the most powerful American launch system in operation is the Titan III family of vehicles, the IIIB, IIIC, IIID, and IIIE. Following the development of the Titan II ICBM, the Air Force pursued development of a space launch vehicle using this powerful booster as the first stage. The first Titan IIIA development vehicle—consisting of a modified Titan II with a third "Transtage" flew on September 1, 1964. Titan IIIB, similar to the IIIA, but with an Agena upper stage, followed shortly thereafter, and went through a mini-growth cycle itself, reminiscent of the Delta experience. The most notable members of the Titan III family, however, are the IIIC, IIID, and IIIE. These combine the basic Titan III with two large solid-fuel strap-on engines giving it a total thrust at launch (when the two strap-ons and the sustainer Titan are firing) of nearly three million pounds of thrust. Titan III boosters were originally developed for the purpose of lofting military payloads into space, such as sophisticated intelligence-gathering satellites. However, the Titan III family has also proven beneficial to the civilian space program, launching such payloads as the Mars Viking landers (which relied on a Titan IIIE-Centaur combination).

Another mid-to-late 1960s upper stage development involved the Boeing Burner II, initiated in 1965 by the Air Force as a means of developing an upper stage to fill the gap between the small Scout and the larger Thor-Delta. The Burner II family utilize solid propellants, and the stage is compatible with Thor, Atlas, and Titan III boosters, as well as with such other upper stages as Agena, Delta, Centaur, and Transtage.[16]

Generally, the United States has been very successful in developing whole families of upper stages for use with modified ICBM and IRBM vehicles. Indeed, in the entire survey of upper stage development, only one disappointment—the Able—is readily discernable. Able, another Vanguard derivative designed for use as a second stage with Atlas and Thor, enjoyed limited success boosting some satellites into orbit when combined with the Thor. But the Atlas-Able combination failed to achieve a single successful launch in three attempts in 1959-1960. Another Atlas-Able exploded on the pad during tests.[10]

It is worth noting that the same patterns that have appeared in America's booster programs have appeared in foreign ones as well: simplicity, building upon existing hardware, seeking low maintenance/low life-cycle cost vehicles rather than highly specialized single-mission launch vehicles. This is especially true in the Soviet Union, where the Soviets, like their American counterparts, drew quickly upon IRBM and ICBM technology to develop advanced sounding rockets and space launch vehicles. A recent case from the People's Republic of China demonstrates that this American-Soviet model holds for the Third World, as well. During the all-too-brief period of Sino-Soviet cooperation in science and technology, the Chinese were introduced to the Soviet SS-3 surface-to-surface missile, using the standard Soviet propellants of liquid oxygen and RP-1 (a petroleum based fuel.)

Following the well-publicized split from Soviet "hegemony," the PRC's small cadres of rocket technologists used the SS-3 as the starting point for their own missile and space launcher program. The sequential derivatives were, apparently, as follows:[36]

CSS-1: a lengthened SS-3 derivative, deployed in 1966 with a 20 kt. nuclear warhead; range of 750 miles.

CSS-2: 1,550 mile range IRBM with a 1 megaton warhead, single stage burning UDMH and nitrogen tetroxide, deployed in 1971. Furnished first stage for subsequent CSS-3 and CSL-1.

CSS-3: Two-stage ICBM, 4,350 mile

Figure 10. Space Shuttle

range, 1 megaton + warhead.

CSL-1 "Long March 1": CSS-2 first stage, CSS-3 second stage, new third stage. Launched China's first satellites in 1970-71.

CSS-X-4: Two-stage 5,000 mile+ ICBM with 5 megaton warhead; not operational, but inspired the CSL-2 space booster.

CSL-2 (FB-1): Has launched all Chinese satellites since 1975. UDMH-nitrogen tetroxide propellants. Can place a 2,645 lb. payload into a 93 mile orbit.

CSL-X-3: A proposed launch system using the first and second stages of the CSL-2 joined with a new high-energy liquid hydrogen upper stage.

A direct comparison between American and Chinese launch vehicles can be made, as indicated below:

Chinese Vehicle	Comparable U.S. Vehicle
CSS-1	Redstone
CSS-2	Thor
CSS-3	Thor-Delta
CSL-1	Advanced Thor-Delta
CSS-X-4	Titan II-warhead
CSL-2	Titan II-satellite payload
CSL-X-3	Titan III-Centaur

Thus, in the Chinese space program, as well as the American and Soviet, innovation and adaption rather than invention and specialization mark the actions of space technologists.

The Space Shuttle: A New Beginning?

When NASA undertook its initial conceptualizations for what eventually evolved into the present-day Space Shuttle, mission planners envisioned the winged shuttle—or a lifting body configuration—boosting into earth orbit to resupply an orbiting space station. Following the collapse of NASA's grandiose post-Apollo plans, the Shuttle, rather then being simply an aspect of those plans, now became the cornerstone of NASA's entire manned spacecraft effort.

This caused, of course, a complete change in the arguments advanced in support of such a craft. Rather than now being a logistical spacecraft, the Shuttle became a manned launch vehicle that would, it was hoped, obviate the need for such launch vehicles as the Delta. The "expendables" would give way to the refurbishable Shuttle. Its total lift-off thrust of 6,925,000 lbs (almost in the Saturn V class) would enable NASA to place a payload weighing 65,000 lbs into orbit 115 miles above the earth. The Shuttle could undertake limited scientific research on its own during short-term orbital missions, and could place and retrieve satellites in space. NASA anticipated that the Shuttle would complete its first orbital flight tests before the end of the 1970s, and would be fully operational by the early 1980s.[15]

Now, however, with the Space Shuttle well behind schedule, and with the heavy projected need for placing satellites into orbit (particularly geosynchronous communications satellites), it is evident that the Shuttle alone cannot be expected to meet user demands. From the perspective of 1980, it is obvious that the Shuttle itself will not replace the expendable "throwaway" launchers such as the Delta. Throughout the 1980s and into the 1990s, the Shuttle at best will supplement, not supplant, older expendable launch systems. Thus, a considerable portion of the booster future rests with vehicles such as the Delta, Scout, and the European Ariane and OTRAG.[5]

What then, does the future hold? Plans are already underway for utilizing Shuttle-carried stages to further enhance its projected effectiveness as a launch system. Insofar as expendable vehicles are concerned, they are, without question, far from being obsolete. It can be expected that as the demands for placing satellites and payloads in space grow, further adaptions of existing launch vehicles and military systems will take place, in part because of the proven track record of success using this approach, but also because developing single-task "mission dedicated" launch vehicles for a specific kind of mission is simply too expensive and time consuming. (Thus, Minuteman may eventually have its day in space, for example.) Like the DC-3, the already near-legendary boosters of the early space age can be expected to fly on for quite a while yet.

REFERENCES

1. Armacost, Michael H. *The Politics of Weapons Innovation: The Thor-Jupiter Controversy* (New York: Columbia University Press, 1969), pp. 54-72, 84-110, 138-155, 166-179.

2. Bilstein, Roger E. "The Saturn Launch Vehicle Family," in Richard P. Hallion and Tom D. Crouch, eds., *Apollo: Ten Years Since Tranquility Base* (Washington, D.C.: Smithsonian Institution Press, 1979), pp. 115-123.

3. Beard, Edmund. *Developing the ICBM: A Study in Bureaucratic Politics* (New York: Columbia University Press, 1976), especially chapters 2, 3, 5, and 6.

4. Blain, J.C.D. *The End of an Era in Space Exploration: From International Rivalry to International Cooperation* (San Diego: American Astronautical Society, 1976), p. 35.

5. Covault, Craig. "Payloads to Saturate Launch Capacity," *Aviation Week and Space Technology* (August 18, 1980).

6. Debus, Kurt. "From A-4 to Explorer 1," paper presented to the 7th International History of Astronautics Symposium, XXIVth Congress of the International Astronautical Federation, Baku, October 1973.

7. Debus, Kurt. "High Altitude Research with V-2 Rockets," *Proceedings of the American Philosophical Society* (Dec. 1947), pp. 430-446.

8. Del Papa, E. Michael and Sheldon A. Goldberg. *Strategic Air Command Missile Chronology, 1939-1973* (Offutt AFB, Neb.: Headquarters, Strategic Air Command, 2 Sept. 1975), n.p.

9. Dornberger, Walter. *V-2* (New York: Viking, 1954).

10. Ezell, Linda N. Unpublished mss., *NASA Historical Data,* "Launch Vehicles," pp. 53 and 101. Copy on file with the NASA Historical Office, Washington, D.C.

11. Green, Constance McLaughlin and Milton Lomask. *Vanguard: A History* (Washington: Smithsonian Institution Press, 1971).

12. Greenwood, John T. "The Air Force Ballistic Missile and Space Program, 1954-1974," *Aerospace Historian* (December 1974), pp. 190-191.

13. Hacker, Barton C. and James M. Grimwood. *On the Shoulders of Titans: A History of Project Gemini* (Washington, D.C.: NASA, 1978), pp. 706-720.

14. Hagen, John P. "The Viking and Vanguard," in Emme, ed., *The History of Rocket Technology: Essays on Research, Development, and Utility* (Detroit: Wayne State University Press, 1964), pp. 122-141.

15. Hallion, Richard P. "The Antecedents of the Space Shuttle," 13th International History of Astronautics Symposium, XXXth Congress of the International Astronautical Federation, Munich, September 1979.

16. Hazard, A.B. and Heinz F. Gehlhaar. *U.S. Space Launch Systems,* Report No. NDDA-R-20-72-2 (Los Angeles: Navy Space Systems Activity, March 1973), pp. 5-49—5-98.

17. Lasby, Clarence G. *Project Paperclip: German Scientists and the Cold War* (New York: Atheneum, 1971).

18. Malina, Frank J. "Is the Sky the Limit?" *Army Ordnance* (July-August 1946), p. 45.

19. Malina, Frank J. "Origins and First Decade of the Jet Propulsion Laboratory," in Emme, ed., *The History of Rocket Technology,* p. 66.

20. Malina, Frank J. "America's First Long-Range Missile and Space Exploration Program: The ORDCIT Project of the Jet Propulsion Laboratory, 1943-1946: A Memoir," 5th International History of Astronautics Symposium, XXIInd Congress of the International Astronautical Federation, Brussels, September 1971.

21. Ordway, Frederick I., III and Mitchell R. Sharpe. *The Rocket Team* (New York: Crowell, 1979).

22. Pendray, G. Edward. "Pioneer Rocket Development in the United States," in Eugene M. Emme, ed., *The History of Rocket Technology,* pp. 19-20.

23. Perry, Robert L. "The Atlas, Thor, Titan, and Minuteman," in Emme, ed., *The History of Rocket Technology,* p. 150.

24. Perry, Robert L. *System Development Strategies: A Comparative Study of Doctrine, Technology, and Organization in the USAF Ballistic and Cruise Missile Programs, 1950-1960,* Rand Memorandum RM-4853-PR (Santa Monica: The Rand Corporation, August 1966).

25. Pickering, William H. with James H. Wilson. "Countdown to Space Exploration: A Memoir of the Jet Propulsion Laboratory, 1944-1958," 6th International History of Astronautics Symposium, XXIIIrd Congress of the International Astronautical Federation, Vienna, October 1972.

26. Rosen, Milton W. "The Viking Rocket: A Memoir," 6th International History of Astronautics Symposium, XXIIIrd Congress of the International Astronautical Federation, Vienna, October 1972.

27. Rosen, Milton, W. *The Viking Rocket Story* (New York: Harper, 1955).

28. Rosen, Milton W. "What Have We Learned from Vanguard?" *Astronautics* (April 1959).

29. Shortal, Joseph A. *A New Dimension: Wallops Island Flight Test Range—The First Fifteen Years* (Wash. D.C.: NASA, 1978), pp. 706-720.

30. Sloop, John L. *Liquid Hydrogen as a Propulsion Fuel, 1945-1959* (Washington, D.C.: NASA, 1978), pp. 187-203.

31. Swenson, Loyd S., James M. Grimwood and Charles C. Alexander. *This New Ocean: A History of Project Mercury* (Washington, D.C.: NASA, 1966).

32. von Braun, Wernher. "The Redstone, Jupiter, and Juno," in Emme, ed., *The History of Rocket Technology,* pp. 107-121.

33. von Kármán, Theodore. Review of Calvin M. Bolster's "Assisted Take-Off of Aircraft," *Journal of the American Rocket Society* (June 1951), pp. 92-93.

34. von Kármán, Theodore with Lee Edson. *The Wind and Beyond: Theodore von Kármán, Pioneer in Aviation and Pathfinder in Space* (Boston: Little, Brown, 1967).

35. Winter, Frank H. *Prelude to the Space Age: Rocket Societies 1924-1940* (Washington, D.C.: Smithsonian Institution Press, 1981).

36. "Chinese Develop Series of Missile, Satellite Launchers," *Aviation Week & Space Technology* (August 25, 1980).

SPACE SCIENCE FOR APPLICATIONS: THE HISTORY OF LANDSAT

PAMELA E. MACK

NASA's Landsat* satellites collect data about features of our planet's surface as seen from space for the study and management of resources. The first satellite was launched in 1972, followed by two more in 1975 and 1978. A fourth and probably a fifth are planned for the mid-1980s. The satellites radio data to earth, where it is processed and printed as pictures or analyzed by a computer. The data can be used for many purposes: for example, to detect large geological features associated with oil and minerals, to measure the areas planted in different crops to help predict harvests, to monitor water distribution and snow cover to predict flooding, and to make maps of land use. Figure 1

*The satellite was originally named Earth Resources Technology Satellite (ERTS), but the name was changed retroactively to Landsat in 1975.

Pamela E. Mack is a predoctoral fellow at the National Air and Space Museum, writing her dissertation for the University of Pennsylvania on the history of the Landsat Earth Resources Satellite.

gives an idea of how these things can be learned: the edge of the white snow is distinct, water appears black, and the rectangular patterns of cultivated fields are visible in some of the valleys. Scientists have shown that many applications are technically feasible, although these are not yet used to their full potential for routine resource management.

The development of Landsat shows the many forces acting on a technological project in the federal government, from interagency rivalry to funding problems to the need to satisfy many different users. In addition to the scientific problems of developing the potential applications, the Landsat project suffered from the political and managerial problems of "applications" satellites. The large range of potential uses required compromises, resulting in a system matching no one's original intentions. Landsat provides a good case study of an experimental, and in part scientific, program to develop an area of practical application of space technology. It thus raises questions of interest to historians of space science, and also a new set of

questions about the balance between the experimental aspects of the project and the ultimate operational goals.

Experimental goals included scientific experiments in the use of the data and the test of one of many possible satellite systems. Such an experimental system is built only to be adequate to show its worth, not for continuing or efficient use. Landsat was a test of one possible system for gathering earth resource data, a test of what such data might be useful for, and a test of practical arrangements for using the data. Satellites are so expensive and take so long to build that tests of many phases of research and development must be conducted at once. The operational or practical use of the satellite can also encroach on the experimental program, as in the case of the first TIROS weather satellite, which was planned as a scientific experiment but was used for practical forecasting starting two weeks after its launch.[9]

An operational satellite system is one used for the routine execution of a job, like crop forecasting to meet the needs of a specific user. NASA sponsored communications and meteorological satellites through the phases from research and development up to operation. By law, NASA's mission is research, so the management of the operational system is turned over to another agency, as in the case of weather satellites, or to a new institution like COMSAT. Landsat has in 1980 still not reached the operational stage, but its data is already put to operational use in some programs.

NASA, the user agencies, and the Bureau of the Budget (since 1970 the Office of Management and Budget) have participated in the project. NASA had responsibility, despite some challenges, for developing a satellite system to observe the earth's resources. The space agency undertook the project in cooperation with some of the federal agen-

Figure 1. Landsat picture of the area around Grand Teton National Park in Wyoming and Idaho. The Grand Teton range is at the top center (NASA photo)

cies that would ultimately use the data from the satellite: the Department of the Interior, the Department of Agriculture, and the National Oceanic and Atmospheric Administration.* Among these user agencies, Interior took the lead in dealing with NASA. In addition to the agencies, users of the data included scientists, industry, state and local governments, and governments and organizations outside the United States. The Bureau of the Budget (BoB) had a major influence on the satellite project by its control of the budgets of NASA and the user agencies.

The history of the Landsat project divides into three phases, each characterized by a dominant problem. In the first phase, from 1964 to 1967, the challenge was to develop interagency cooperation and to achieve consensus on basic plans for the satellite. In the second, from 1968 to 1971, the cooperating agencies had to persuade the Bureau of the Budget to provide funding for the project. In the third, since 1972, the challenge to NASA has been to encourage applications of the Landsat data and plan the shift from an experimental to an operational program. The tension between experimental and operational goals has run through all these phases, and can provide the basis for an understanding of the interaction of technological and political systems.

I. Interagency Politics

The many uses of the Landsat data engendered cooperation between NASA, the Department of the Interior, the Department of Agriculture, and the National Oceanic and Atmospheric Administration. To cooperate, the agencies first had to work out their respective roles and learn to understand each other's areas of expertise and responsi-

*In 1965 the Weather Bureau and the Coast and Geodetic Survey became part of the newly founded Environmental Science Services Administration. This, in 1970, became part of the new National Oceanic and Atmospheric Administration, which also subsumed the satellite oceanography program of the Naval Oceanographic Office.

bility. They had to compromise on their requirements so that the system would satisfy them all, since BoB would not allow each one to build a satellite for its own purposes.

The Department of Agriculture and the Department of the Interior had employed aerial photography for mapping for many years, but the idea of observing the earth from space for civilian purposes probably grew out of military reconnaissance from satellites. Aerial reconnaissance in WWII produced a community of experts in photogrammetry (surveying and mapping from photographs) and stimulated research on sensors other than photographic cameras. Photogrammetry formed the core of the new field of remote sensing, which was defined to include any gathering of information without touching the object. The field remained, however, primarily oriented toward the military. In 1962, at the first Symposium on Remote Sensing of Environment at the University of Michigan, one of the participants wrote: "Modern aircraft and earth satellites now make it possible to survey the earth from a vantage point heretofore unavailable. Surveillance systems integrated into these vehicles make possible the rapid accumulation of geophysical data over large areas of the terrestrial environment."[11] Circumstantial evidence in the conference proceedings suggests that some of these scientists were already using photographs taken by reconnaissance satellites for civilian investigations of geology and land use. The desire of these scientists to continue this research on an unclassified basis was probably the first source of interest in a civilian satellite.

In 1964, at the urging of Peter Badgley, a geologist in the Apollo program, NASA created an office to study the possibility of investigating earth resources from space. NASA immediately included the potential user agencies by transfering $100,000 to the Department of the Interior and lesser amounts to the Department of Agriculture and the Naval Oceanographic Office for studies of how such a satellite could help

them carry out their missions.[5] The space agency also set up an aircraft program at the Manned Spacecraft Center (now the Johnson Space Center) in Houston to test sensors and develop techniques for using the data. Much of this research concentrated on the scientific problems involved in extracting useful information from the vast amounts of data that a satellite would gather. For example, in order to determine the meaning of a particular color on an image or reading on an instrument, a researcher walked through the field making direct observations of crop variety and health (this is one form of what is called collecting ground truth) while an airplane flew over with the sensor. Once this was done, the data from the sensor could, in theory, be interpreted alone.

NASA continued to fund its own research and that of other agencies in 1965 and 1966, but did not decide what sort of satellite to fly or seek approval and funding for it. The space agency opposed an early launch, favored by Interior, because NASA wanted to test a wide variety of sensors from aircraft. NASA then planned to launch a satellite or manned mission (in the planned Apollo Applications series that later became Skylab) including all the most promising sensors so that scientists could compare the results of different sensors.[13] Other factors discouraging planning for an unmanned satellite probably included the need for practical applications to justify a continuing manned program and fears that a civilian satellite producing pictures of the earth would threaten national security by calling attention to all reconnaissance satellites. NASA planned an extensive research program to develop the best possible operational system and to allow any political problems to be cautiously resolved.

Impatient with the delay, the Department of the Interior decided to take action.* On September 21, 1966, Secretary Stewart Udall announced an Interior program to launch its own satellite.[26] In fact, the people behind this move, which created the EROS (Earth Resource Observation Satellite) program, did not expect that Interior would be allowed funding for a satellite program independent of NASA. Rather, they believed that their announcement would make NASA stop procrastinating and show the Department of the Interior as a forward-looking agency.[22] The EROS proposal called for the launch of a simple operational earth resources satellite, carrying a sensor similar to a television camera.[26] The Department of the Interior believed that this sensor would be immediately useful and did not want to wait for NASA to test other, more complex, sensors. In order to be funded, Interior, as an operating agency, had to present the program as an operational tool, while NASA cast it by necessity as a research project.

The EROS announcement shocked almost everyone at NASA. The responsibility for an operational earth resources system might properly be assiged to the Department of the Interior or the Department of Agriculture (or to the National Oceanic and Atmospheric Administration as occurred finally in late 1979), but NASA assumed that that decision would follow an experimental satellite system which would clearly be the responsibility of NASA as the research agency. Skipping that step might imply that NASA's role as a research agency was superfluous, that the agency that would use the operational system could simply contract directly with industry for research and development of the satellites it wanted. Administrator James Webb took NASA's objections to President Lyndon Johnson, who agreed that NASA should conduct research and development before an operational satellite could be approved.[23] The EROS program survived, but only as a focus for Interior's cooperation with NASA under the

space agency's leadership.

NASA and the user agencies worked together on the project, but they hotly debated the design and timetable for an operational satellite system during the discussion of an experimental satellite plan. The EROS group sent specifications for the satellite they wanted to NASA in October 1966. They called for an operational system by the end of 1969 providing repetitive observations of the entire surface of the earth by a sensor producing pictures with a resolution of 100 to 200 feet.[15] NASA's Deputy Administrator replied in April 1967 that many aspects of the specifications were "just at or beyond the state of the art" and therefore required an orderly program of experimentation and development to avoid excessive costs.[24] The Department of the Interior then offered to participate in the funding of the first satellite if it was a modification of the TIROS weather satellite,[16] but nothing came of this offer.*

A number of areas of disagreement among the agencies remained. The Earth Resources Office at NASA wanted a more elaborate experiment than could be mounted on a TIROS. The Department of Agriculture wanted a different sensor, having learned from aircraft experiments that observations in the infrared and more precise differentiation of colors were needed to distinguish crops and detect the effects of disease and drought. By June 1967 NASA had persuaded the EROS group to leave the concept of an operations system "for some distant time schedule."[20] In September 1967, NASA's Assistant Administrator for Policy complained to his boss that the user agencies still did not understand the need for research and development.[25]

While NASA wanted a large satellite to test a variety of sensors, the different user

---

*Two people, William Fischer and Charles Robinove, who had been working on NASA-sponsored research at the Department of the Interior, decided that Interior was ready for an operational satellite. They convinced the Director of the Geological Survey, William Pecora, who in turn convinced Udall.[22]

*Interior was not the only agency that preferred the TIROS design. TIROS was the experimental satellite selected by the Weather Bureau for their operational system instead of the more expensive and more elaborate Nimbus satellite design that, with modifications, became Landsat.[4]

agencies made mutually conflicting demands for small satellites with simple sensors to serve their different missions. The EROS group at the Department of the Interior advocated a kind of television camera called a Return Beam Vidicon. The RBV used a shutter to expose a light-sensitive plate and then scanned the plate with an electron beam to capture the image on video tape or radio it to the ground. Three RBV's with different colored filters allowed the reconstruction of color pictures. The Department of Agriculture group wanted a sensor that would give better spectral resolution—more precise colors—at the expense of having more distortion which would interfere with map making (desired by the Geological Survey, part of the Department of the Interior). This Multi-Spectral Scanner used a mirror to scan the scene on the ground one line at a time, reflecting the light onto a series of detectors (photoelectric cells) sensitive to four different spectral regions. Users interested in hydrology (water science and management) wanted the satellite to receive data from sensors on the ground in remote areas, such as water level gauges in mountain streams, and relay it back to the ground with the satellite data. NASA selected these two sensors and the data relay system for Landsat. The requirements of the oceanography community, represented by the Naval Oceanographic Office and the National Oceanic and Atmospheric Administration, were also discussed, but NASA decided to plan a separate satellite for oceanography (Seasat, launched in 1978) because too many of its requirements contradicted those of the other disciplines.

The controversies between the agencies in the first few years of the earth resources satellite program resulted from conflicting and overlapping interests in the specification, design, and operation of such a satellite. As political scientist Don Price has pointed out, the growth of the federal role in science and technology has blurred the boundaries between the executive agencies

by creating new interrelations and dependencies.[21] This blurring makes cooperation more necessary because an agency can no longer be confident of its hold on its domain. In the case at hand, the domains could not be divided by assigning NASA responsibility for the experimental program and giving another agency the operational program. The user agencies demanded control from the start, because the experimental program would inevitably shape the operational program.

II. Funding

What finally ended the lack of cooperation between NASA and the Department of the Interior was a common antagonist, the Bureau of the Budget. By the time NASA requested funding for Landsat, in the fall of 1967 for Fiscal Year (FY) 1969, rising costs of the manned space program and declining enthusiasm for space had led to close scrutiny of NASA's budget. The Bureau of the Budget opposed the Landsat project from the start and BoB's successor, the Office of Management and Budget, continues to do so to this day, pointing to the slow development of operational use of the satellite data as justification.

The Bureau of the Budget tried to prevent NASA from starting development of the Landsat system. NASA supported research on Landsat from its general research funds until the FY 1969 budget, then included Landsat development of the proposed budget. When BoB received this budget in the fall of 1967, the bureau cut Landsat from the budget entirely. NASA Administrator Webb took his case to President Johnson, who approved funding for Landsat, but only for cost-benefit studies, not for development. In 1968 BoB rejected Landsat again for FY 1970, but this time the appeal to the President produced some money for development, allowing NASA to award study contracts to Thompson Ramo Wooldridge, Inc., (TRW) and General Electric in October 1969. The Department of the

Interior had less success for FY 1970; BoB severely cut back the funds for a proposed center to distribute Landsat data. Congress increased the appropriation to $4.1 million, but BoB impounded $3 million of that.[5] BoB totally eliminated Interior's budget for EROS in the Fall 1969 decisions of the FY 1971 budget, and cut back NASA's allotment for Landsat by postponing the launch of the first satellite until after 1972. By this point Landsat was so well established in the press that the agencies could argue that "The abrupt [apparent] change in Administration policy proposed by BoB cannot help but become a major embarrassment to the Administration."[12] NASA regained funds to allow a 1972 launch, but the Department of the Interior did not receive enough money to put the Data Center back on schedule. Resistance by the Office of Management and Budget has continued to delay subsequent Landsat satellites.

BoB's opposition had at least four causes. First was the desire to hold down the federal budget, which meant that all new projects came under careful scrutiny. Second, wide experience with the problems of new projects probably led to the suspicion that NASA promised benefits without a full understanding of the difficulty of moving a new idea into practical use. Third, the BoB staff, a number of whom were members of the military or the CIA detailed to BoB, worried about the risks of calling attention to reconnaissance satellites, which were key to national security.[6] Fourth, BoB questioned whether satellites could perform earth resources surveying as well as aircraft. A RAND Corporation scientist, Amrom Katz (who had friends at BoB),[2] argued that the resolution planned for Landsat data would make it useless and that an aircraft program could do a better job for less money.[14] Some staffers at BoB had experience with reconnaissance satellites, from which they had learned the advantages of fine resolution images. This feeling was reinforced when NASA proposed to provide coarser resolu-

tion than originally requested by the user agencies.[18] In fact, NASA had convinced the user agencies that multispectral (color) data, even at coarse resolution, could provide more information than was at first apparent to users accustomed to fine-resolution aircraft photographs.

Because of these concerns, BoB, when finally forced by the decisions of the President to fund Landsat, put restrictions on the project that proved to be crucial to the development of the technology. Since NASA argued that the goal of this satellite, or at least of the eventual operational system, was to improve the management of resources, BoB required NASA to show by cost-benefit analysis that the benefits claimed could not be achieved at a lower cost with some other system, such as an aircraft program. BoB argued that the security risk should not be taken until the benefits of a civilian satellite clearly outweighed the risk, while NASA argued that it was impossible to predict the benefits of the data until after an experimental satellite program had allowed scientists to see what they could do with it. Because of BoB's doubts about the eventual usefulness of the satellite project, the bureau required NASA to commit itself to no plans and to buy no equipment for an operational system until an experimental system had proved its worth.[17] In other words, NASA was trapped in a circular situation. It had to conduct only an experimental, scientific project, but could justify it only on the grounds of its eventual practical uses, not on the grounds of the scientific research that could be done with the satellite data.[7]

BoB insisted that Landsat be an experimental system to a point that frustrated NASA. The Landsat project was an experiment in data use, and therefore required that the data be routinely and promptly available. BoB held, however, that the data-processing system had to meet only BoB's standards of reliability for an experiment,[17] which resulted in an undependable system producing an irregular flow of data.

The limitation of funds affected not only the reliability of the system but also the type of technology used in the data processing system. The sensors on Landsat, chosen to meet the requirements of the user agencies, radioed down to earth more than 15 million bits of data per second, far more than any other civilian satellite at the time. This data had to be corrected, recorded on computer tape or printed on photographic film, and copied for a large number of users. The two companies, TRW and GE, that received study contracts for Landsat in 1969 proposed two very different data-processing systems. TRW planned to use a large digital computer to perform the corrections and special-purpose processing and store and retrieve the data, but the computer technology to process such a large amount of data was not fully developed and sure to be expensive. GE chose to develop a less expensive system by using a combination of analog and digital technology:* corrections (calculated by a digital computer) were applied to the data during the process of printing it on photographic film, and photographic techniques were used to store and retrieve the data and provide special-purpose processing. NASA chose GE's proposal in July 1970 because it was less expensive in the short run, although many people at NASA realized that future uses of the full potential of the Landsat data would require the extra capabilities of a digital system.

The Bureau of the Budget provided one of the many sets of conflicting demands placed on the satellite. BoB not only gave NASA little money for Landsat, it also specified that the money be used only on an experimental project and further shaped the project by demanding cost-benefit analyses for justification. Tight money meant that innovation had to be more conservative; NASA could not afford to take the risk of using new technology that might not be fully developed by the time it was needed.

III. Applications

After the launch of the first satellite in 1972, exploring and encouraging the use of the data became the critical problems for NASA. There were three groups of potential users: scientists, federal agencies with resource management responsibilities, and non-federal operational users such as state and local governments and private companies. In order to prove to BoB the value of the satellite, NASA had to develop a strategy for involving these users and promoting the most beneficial applications of the data from the satellite.

From the start, the user agencies developed practical techniques for the use of Landsat data for their own purposes. The early grants of funds from NASA* allowed the user agencies to conduct, and to let contracts to universities for, the basic research testing sensors and collecting ground truth necessary to understand the data from the satellite. The Department of the Interior encouraged many small projects in different branches to investigate the use of satellite data for different purposes, while the Department of Agriculture concentrated its efforts on a few large projects to investigate crop identification and monitoring of crop conditions for agricultural surveys.

NASA at first expected the user agencies to develop applications for the data, but soon realized that the user agencies had no authority to study applications outside their own mission. To encourage the widest variety of possible uses for Landsat data, NASA in 1970 requested proposals for experiments from scientists throughout the

---

*An analog system handles the data as continuous shades of grey, such as on a photograph. A digital system handles the data by assigning a number describing the discrete shade of grey (perhaps on a scale from 1 to 64) at each point on the image.

*The Office of Management and Budget stopped NASA's transfers of money to the user agencies in 1972 on the grounds that the user agencies should show their enthusiasm for the project by spending their own money on it.[5]

Figure 2. Landsat 1 (NASA photo)

world. The successful applicants (the majority of them scientists at universities) were called principal investigators, although they did not participate in the design of the instruments or have first rights to the data like principal investigators for NASA's scientific satellites. NASA chose more than 300 Landsat principal investigators to receive data, and in many cases funding, in return for reporting to NASA the results of their experiments.

NASA also decided to release data to the public immediately. This contradicted the usual agency policy of allowing principal investigators exclusive rights to their data for a few months to allow the scientists who had invested time and effort planning the experiment to publish the first results. In the case of Landsat, however, some information could potentially be used to make money, which raised the possibility that some experimenter would exploit an advantage gained from the data.* The easiest way to prevent this was to release the data immediately to the public, so that anyone could theoretically gain the same advantage.[1] This policy of completely open data has also been the foundation of a successful program encouraging the international use of Landsat data. The public can purchase Landsat pictures (and computer tapes) from the EROS Data Center, Sioux Falls, South Dakota 57198.

The first experiments with Landsat data provided promising results. Geologists found that the data revealed many new large-scale features, particularly faults and other linear structures, although it could not be used to classify rocks, as some had hoped. Agricultural scientists discovered that crops could be identified and their areas measured for small regions, but predicting harvests for the country and the world would require

extensive study of regional variations and a tremendous amount of work on computer classification and analysis. Hydrologists found the Landsat images more useful than expected for mapping water areas, both normal and flooded, and monitoring snow cover and glaciers.[10]

The Landsat managers at NASA learned, however, that making the data easily available and sponsoring scientific experiments was not enough to insure the fullest possible exploitation of Landsat data. They therefore established a variety of "technology transfer" programs for resource managers in state and local governments (industry use, mostly for oil and mineral exploration, has developed with less assistance). A Regional Applications Program with offices dedicated to technology transfer served each region of the United States (the first office was set up in 1970), but this program had little effect until after a reorganization in 1977. NASA started an Applications System Verification and Transfer Program in 1975, which has concentrated on establishing prototype full-scale operational systems instead of small experiments.[19]

State officials registered many complaints about the technology transfer program. They charged that NASA was not putting

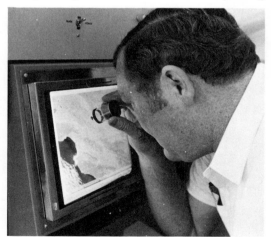

Figure 3. A NASA scientist, Dr. Arch Park, examining a Landsat image on a color-enhancing machine (NASA photo)

---

*For example, if a Landsat principal investigator discovered evidence of oil on land thought to be worthless and obtained the land at a low price because no one else had the data to learn about the oil, the owner of the land could argue that NASA had given the investigator an unfair advantage.

enough effort into technology transfer, particularly in training state people and providing technical assistance and seed money to reduce the risk to the states of developing a capability to use the data and then finding it not worthwhile. Describing another problem, Governor Reubin Askew of Florida wrote in 1978, "This interesting new technology has not been used regularly by non-Federal interests and agencies because they had no input into the design capabilities of satellites."[19] This expressed a widely held feeling that state and local governments needed to participate in formulating plans and policy for the satellite system before they could use it effectively. Perhaps the most serious barrier to state and local participation, however, was the lack of commitment to an operational system. States were understandably reluctant to invest in a system to use Landsat data while Landsat remained an experimental system that could be canceled or radically changed at any time.

NASA has slowly extracted Landsat from this trap, and the system is gradually becoming operational. President Carter assigned the responsibility for managing operational civilian remote-sensing satellite activities to the National Oceanic and Atmospheric Administration in November 1979. An interim operational system based on Landsat D is planned for the mid-1980s, but a system designed as fully operational is not planned before 1990.[8]

The transition from an experimental technique—identification of corn and soybeans on a Landsat image—to an operational system—predicting entire harvests—is a difficult and poorly understood management problem. NASA had to deal with a particularly complex case in Landsat because of the many unrelated applications and the many users with different interests. In such a case the transition was necessarily slow.

IV. Conclusion

The case of Landsat shows some of the problems of developing new technology in our present era in a government setting. The Landsat system started out from the "push" of a new technology, but it attracted the interest of agencies whose different goals hindered cooperation. One of the most fundamental conflicts between agencies was that between an experimental orientation and a mission or operational orientation. This conflict shows not only the problems of managing a project besieged by contradictory demands but also one mechanism by which inter-agency politics affects technological systems.

Landsat derived from innovation by scientists and engineers rather than the pull of a market, but the project soon found itself buffeted in a market of user agencies. The user agencies were buying agency expansion by relating the new product to their missions, and each brought its own goals to Landsat. The program became caught in a difficult situation between the conflicting goals of NASA, the user agencies, and BoB. Working a way out of this situation has required compromises on all sides, resulting in a satellite project different from that intended by any of the agencies.

NASA preferred to stress the experimental aspects of Landsat. Research agencies have an interest in prolonging the experimental portion of any program as long as any questions are unresolved. In the late 1960s and early 1970s, when Landsat was being developed, NASA was planning for the end of the Apollo program and felt the need to find new goals for the space program. Landsat provided an opportunity to answer critics of the space program by proving that it was useful to people on earth. The style of NASA's Office of Space Science and Applications (Applications became a separate office in 1972) was predominantly scientific, with an emphasis on launching series of satellites, each one better than the one before, in continuing research programs. This style had already caused conflict with the Weather Bureau over meteorological satellites because NASA continued its program of experimental satellites until it was satisfied. The Weather Bureau, however, decided that the satellite NASA had chosen as the operational prototype was too complex and expensive, and instead chose to base the operational system on an earlier experimental satellite design (TIROS).[4]

The user agencies for Landsat wanted an operational satellite program as soon as possible. These agencies were judged by how well they performed the missions assigned to them by Congress, so they were anxious to put satellites to work in the routine execution of those missions. NASA could only budget for research and development while the user agencies had to budget for operations. The distance between an experiment, testing feasibility, and routine execution can rarely be covered in one step. Instead, the users needed a technology transfer program based on tests of data utilization—full scale experiments dealing with all the problems of operational use of the data. Only this sort of test would persuade resource managers to integrate the data into their operational systems.

The experimental-operational tension formed a link between interagency politics and a technological system. Science and technology have created new tools that can serve many domains and also a new kind of domain, research. NASA's domain is defined not by a subject area, like agriculture or national resources, but by a research tool, the space vehicle. The domain conflict between NASA and Interior was not between one subject area and another but between the experimental and the operational use of satellites to detect and monitor natural resources. NASA also had to deal with the desire of each user agency for a satellite designed specifically for problems in that agency's domain. The agencies translated their interests in particular domains into goals for the satellite: an operational natural resource system, an operational agricultural system, or an experimental system to test all the possible uses of the new space technology. The political maneuvering among these agencies determined what combinations of goals the satellite project would actually meet, and these goals in turn determined the technology and the configuration of the satellite system. Thus interagency politics affects technological systems by the competition of political interest groups which have each chosen a technology serving their own needs.

The author would like to thank the NASA History Office for financial support of much of her research on Landsat. The opinions I express are not necessarily those of NASA. I would also like to thank Bernie Carlson, Paul Hanle, Thomas Hughes, William Raney, Alex Roland, Bayla Singer, and Monte Wright for advice and criticism.

REFERENCES

1. Allnut, Robert F., letter to Thomas G. Abernethy, 12 Aug. 1969. NASA History Office, folder: "Documentation—Earth Resources."

2. Anders, William A., letter to George M. Low, 24 Feb. 1971, with enclosed memo from David Elliott to George M. Low, "ERTS Resolution." Washington National Records Center, Record Group 255, Accession 77-0677, box 33, folder: "Related Sciences 3 Remote Sensing ER/ERS/EROS/ERTS Jan.-Mar. #1, 1971."

3. Anderson, Clinton P., letter to James C. Fletcher, 14 Oct. 1972. NASA History Office, folder: "Fletcher Correspondence."

4. Chapman, Richard LeRoy. *A Case Study of the U.S. Weather Satellite Program: The Interaction of Science and Politics.* Syracuse University Ph.D. Thesis, 1967.

5. DeNoyer, John M. "Summary of the History, Role and Activities of the Earth Resources Observations Systems (EROS) Program: Contribution to a Congressional Study of U.S. Civilian Space Activities." Prepared for the Library of Congress, Congressional Research Service, at the Request of the Subcommittee on Space Science and Applications, U.S. House of Representatives, July 1978. Provided by William Hemphill.

6. Fischer, William A., interviewed on 8 Aug. 1978 in Reston, Virginia.

7. Fletcher, James C., note to Dr. Low, 30 Oct. 1974. NASA History Office, folder: "Fletcher Correspondence."

8. Frank, Richard A. "Statement by Mr. Richard A. Frank, Administrator, National Oceanic and Atmospheric Administration, on Planning for a Civil Operational Land Remote Sensing Satellite System. Before the Subcommittee on Science, Technology and Space of the Committee on Commerce, Science and Transportation, United States Senate," 26 June 1980. NASA History Office, folder: "Earth Resources 1980-."

9. Freden, Stanley C., Enrico P. Mercanti, and Donald E. Witten, eds. *Symposium on Significant Results Obtained From Earth Resources Technology Satellite – 1.   Volume II: Summary of Results.* Greenbelt, Maryland: Goddard Space Flight Center, May 1973.

10. Freden, Stanley C., and Enrico P. Mercanti, eds. *Symposium on Significant Results Obtained From Earth Resources Technology Satellite – 1.   Volume III: Discipline Summary Reports.* Greenbelt, Maryland: Goddard Space Flight Center, May 1973.

11. Infrared Laboratory, Institute of Science and Technology, University of Michigan. *Proceedings of the First Symposium on Remote Sensing of Environment: 13, 14, 15 February 1962.* Ann Arbor: University of Michigan, March 1962.

12. "Interior Department Appeal of 1971 Budget Allowance: EROS," received at NASA 25 Nov. 1969. Washington National Records Center, Record Group 255, Accession 77-0677, box 33, folder: "Related Sciences 3, ERTS/EROS,NASA/Interior Collaboration."

13. Jaffe, Leonard, and Peter Badgley, "NASA Natural Resources Program," 13 May 1966. NASA History Office, folder: "Documentation—Earth Resources."

14. Katz, Amrom H. "Let Aircraft Make Earth-Resources Surveys," *Astronautics and Aeronautics* 7 (June 1969): 60-68.

15. Luce, Charles F., letter to Robert C. Seamans, Jr., 21 Oct. 1966, with enclosure, "Operational Requirements for Global Resource Survey by Earth-Orbital Satellites: EROS Program." Washington National Records Center, Record Group 255, Accession 77-0677, box 33, folder: "Related Sciences 3, ERTS/EROS, NASA/Interior Collaboration."

16. Luce, Charles F., letter to Robert C. Seamans, Jr., 24 April 1967. Washington National Records Center, Record Group 255, Accession 77-0667, box 33, folder: "Related Sciences 3, ERTS/EROS, NASA/Interior Collaboration."

17. Mayo, Robert P., letter to Walter J. Hickel, 14 April 1970. W. Fischer's Significant Documents file, volume 2, EROS Program Office, Reston, Va.

18. "NASA FY 1971 Budget: Reclama of BoB Tentative Allowance." NASA History Office, folder: "Landsat 1 Documentation."

19. Natural Resource and Environmental Task Force of the Intergovernmental Science, Engineering and Technology Advisory Panel (ISETAP), Office of Science and Technology Policy. "State and Local Government Perspectives on a Landsat Information System," June 1978.

20. Pecora, W.T., memo to the Under Secretary, Department of the Interior, "Status of EROS Program," 15 June 1967. W. Fischer's Significant Document File, volume 1, EROS Program Office, Reston, Va.

21. Price, Don K. *The Scientific Estate.* Cambridge, Mass.: Belknap Press of Harvard University Press, 1965.

22. Robinove, Charles, interviewed on 31 July 1978 in Reston, Virginia.

23. Seamans, Robert C., Jr., letter to Stewart L. Udall, 22 Sept. 1966. NASA History Office, folder: "ERTS Documentation JEW & RCS."

24. Seamans, Robert C., Jr., letter to Charles F. Luce, 7 April 1967, with enclosure, "Specific Comments on October 21 from C.F. Luce to R.C. Seamans," dated 13 Jan. 1967. NASA History Office, folder: "Landsat 1 Documentation."

25. Smart, Jacob E., memo to Dr. Seamans, "Meetings with Representatives of the Departments of Agriculture and Interior Re Earth Resources," 11 Sept. 1967. Washington National Records Center, Record Group 255, Accession 77-0677, box 33, folder: "Related Sciences 3, ERTS/EROS, NASA/Interior Collaboration."

26. Udall, Stewart L., "Earth Resources to Be Studied from Space," 21 Sept. 1966, United States Department of the Interior News Release. EROS Program Office Active Files, folder: "EROS Program—Creation."

CONTINUING HARVEST: THE BROADENING FIELD OF SPACE SCIENCE

HOMER E. NEWELL

The following essay is a republication of Chapter 20 of Homer Newell's recent monograph Beyond the Atmosphere: Early Years of Space Science *(Washington, D.C., GPO, 1980) (NASA SP-4211). We print it here, with the author's permission, because it provides a comprehensive overview of the space science programs that NASA supported in the 1960s and early 1970s.*

As the decade of the 1960s neared its end, space science had become a firmly established activity. While the past had been immensely productive, the future promised much more and thousands of scientists around the world bent to the tasks that lay ahead. A steady stream of results poured into the literature; universities illustrated courses in the earth sciences, physics, and astronomy with exam-

Homer E. Newell was, before retirement in 1973, Associate Administrator of the National Aeronautics and Space Administration. He was responsible for the space-science programs of that agency.

ples and problems from space research, and a few offered courses devoted entirely to space science. For their dissertations graduate students worked with their professors on challenging space science problems. With the loss of that air of novelty and the spectacular that had originally diverted attention from the purposefulness of the researchers, the field had achieved a routineness that equated to respectability among scientists.

Maturity underlay the field's hard-earned respectability. Starting about 1964, in addition to the individual research articles published in the scientific journals, more comprehensive professional treatments of the kind that characterizes an established, active field of research began to appear.[1] It is interesting, for example, to compare the book *Science in Space* published in 1960 with the second edition of *Introduction to Space Science* issued in 1968.[2] The matter-of-fact tone of the latter, which discussed what space science had already done and was doing for numerous disciplines, contrasts with the promotional tone of the former, which could only treat the potential of space science, what rockets and spacecraft might do for various scientific disciplines.

SPACE SCIENCE AS INTEGRATING FORCE

The breadth of the field as it evolved was impressive. Among the disciplines to which space techniques were making important contributions were geodesy, meteorology, atmospheric and ionospheric physics, magnetospheric research, lunar and planetary science, solar studies, galactic astronomy, relativity and cosmology, and a number of the life sciences. The assured role of space science in so many disciplines in the late 1960s was a source of considerable satisfaction to those who had pioneered the field, an ample justification of their early expectations.

But more significant was the strong coherence that had begun to develop among certain groups of space science disciplines. Perhaps the most profound impact of space science in its first decade was that exerted upon the earth sciences. Sounding rockets made it possible to measure atmospheric parameters and incident solar radiations at hitherto inaccessible altitudes and thus to solve problems of the atmosphere and ionosphere not previously tractable. Satellites added a perspective and a precision to geodesy not attainable with purely ground-based techniques. The improved precision laid a foundation for establishing a single worldwide geodetic network essential to cartographers who wished to position different geographic features accurately relative to each other. The new perspective gave clearer insights into the structure and gravitational field of the earth. These examples illustrate one of several ways in which space science was affecting the earth sciences; that is, making it possible to solve a number of previously insoluble problems.

Following James Van Allen's discovery of the earth's radiation belts and the growing realization over the ensuing years that these were but one aspect of a tremendously complex magnetosphere surrounding the earth,

magnetospheric research blossomed into a vigorous new phase of geophysical research. This was a second way in which space science contributed to the earth sciences, opening up new areas of research.

But probably the most significant impact of space methods on geoscience was to exert a powerful integrating influence by breaking the field loose from a preoccupation with a single planet. When spacecraft made it possible to explore and investigate the moon and planets close at hand, among the most applicable techniques were those of the earth sciences, particularly those of geology, geophysics, and geochemistry on the one hand and of meteorology and upper atmospheric research on the other. No longer restricted to only one body of the solar system, scientists could begin to develop comparative planetology. Insights acquired from centuries of terrestrial research could be brought to bear on the investigation of the moon and planets, while new insights acquired from the study of the other planets could be turned back on the earth. Delving more deeply into the subject, one could hope to discern how the evolution of the planets and their satellites from the original solar nebula—it being generally accepted that the bodies of the solar system did originate in the cloud of gas and dust left over from the formation of the sun—could account for their similarities and differences.

The wide range of problems served to draw together workers from a number of disciplines. Astronomers found themselves working with geoscientists who came to dominate the field of planetary studies that had once been the sole purview of the astronomers. Physicists found in the interplanetary medium and planetary magnetospheres a tremendous natural laboratory in which they could study magnetohydrodynamics free from the constraints encountered in the ground-based laboratory. Also known as hydromagnetics, this field was an extension of the discipline of hydrodynamics to fluids that were electrically charged (plasmas), particularly their interactions with embedded and external magnetic fields. The scientific importance of the field stemmed from the realization that immeasurably more of the matter in the universe was in the plasma state than in the solid, liquid, and gaseous states of our everyday experience. An outstanding practical value lay in the fact that magnetohydrodynamics was central to all schemes to develop nuclear fusion as a power source. Physicists also found the opportunity to conduct experiments on the scale of the solar system attractive for the study of relativity, and many of them began to devise definitive tests of the esoteric theories that were in existence. It is safe to say that this interdisciplinary partnership was a valuable stimulation to science in general.

The expanding perspective derived from space science was, in the author's view, the most important contribution of space methods to science in the first decade and a half of NASA's existence. While it was natural for individual scientists to concentrate attention on their individual problems, to those who took the time to assess progress across the board, the growing perspective was clearly evident even in the early years of the program. In a talk before the American Physical Society in April 1965, the author addressed himself to the growing impact of space on geophysics, which even then appeared much as described above.[3] NASA managers in their presentations to the Congress began to emphasize the important perspectives afforded by space science. As a case in point, the Spring 1967 defense of the NASA authorization request for fiscal 1968 described space science as embracing (1) exploration of the solar system and (2) investigation of the universe.[4] Gathering the different space science disciplines into these two areas was not simply a matter of convenience. Rather it reflected a growing recognition of the broadening perspective of the subject, a point that was further developed by Leonard Jaffe and the author in a paper published in *Science* the following July.[5] At the time it was much easier to treat of the impact of space science on the earth sciences, which already offered many examples. While it would probably take a number of decades to achieve a thorough development of the field of comparative planetology, with an appreciable number of missions to the moon and planets behind and more in prospect, the powerful new perspectives available to the geoscientists were quite clear.

As for astronomy—the investigation of the universe—the deeper significance of the impact of space science on the discipline appeared to be unfolding more slowly. To be sure, the most obvious benefit—that of making it possible for the astronomer to observe all wavelengths that reached the top of the atmosphere, instead of being limited to only those that could reach the ground—began to accrue with the earliest sounding rockets that photographed the sun's spectrum in the hitherto hidden ultraviolet wavelengths. This benefit grew steadily with each additional sounding rocket or satellite providing observations of the sun and galaxy in ultraviolet, X-ray, and gamma-ray wavelengths. The value of these previously unobtainable data was inestimable. But in the long run, a deeper, more significant impact of space methods on astronomy could be expected, as Prof. Leo Goldberg and others pointed out: the advent of a much more powerful means of working between theory and experiment than had ever existed before.

At one time the author tried to persuade the House Subcommittee on Space Science and Applications that, as far as the origin and evolution of natural objects were concerned, the scientist knew more about the stars than about the earth. The statement was intentionally phrased in a provocative fashion to get attention, which it did. The Congressmen reacted immediately in disbelief, and it took quite a bit of discussion to develop the point, which went as follows.

Certainly men living on the earth, as they

do, had been able to amass volumes and volumes of data on the earth's atmosphere, oceans, rocks, and minerals of a kind and in a detail that could not be assembled for a remote star. But, when it came to the question of just when, where, and how the earth formed and began to evolve many billions of years ago, the scientist was limited to a study of just one planet—the earth itself. From an investigation of that one body and whatever he could decipher of its origin and evolution, he had to try to discern the general processes that entered into the birth and evolution of planets in general. Only in such a broad context could the scientist feel satisfied that he really understood any individual case. Having only the earth to study, he was greatly hampered.

For the stars, however, the astronomer had the galaxy containing 100 billion stars to observe, and billions of other galaxies of comparable size. In that vast array the astronomer could find, for any object he might want to study, examples at any stage of evolution from birth to demise. With such a display before him in the heavens, the astronomer could proceed to develop a theory of stellar formation and evolution and test the theory against what he observed. In such an interplay between theory and observation the theorists did develop a remarkable explanation of the birth, evolution, and demise of stars.[6] So, in this sense, the astronomer could claim to understand more about the stars than the earth scientist did about the earth.

But there was a shortcoming in this theoretical process. The theory was based on observations of those wavelengths that could reach the ground—mostly visible, with a little in the ultraviolet and infrared, and after World War II critical observations in some radio wavelengths. Yet that very theory predicted that vitally important stellar phenomena would be manifested in the emission of wavelengths that the astronomer could not yet see. The early formation of a star from a cloud of gas and dust that was beginning to aggregate into a ball would be revealed primarily in the infrared as gravitational pressures caused the material to heat up. At the other end of the spectrum very hot stars would be emitting mostly in wavelengths shorter than the visible, presumably mostly in the ultraviolet. Little attention was paid to X-rays or gamma rays, yet among the most important discoveries of space science have been X-ray emissions from the sun and more than a hundred stellar sources.[7] Here is where the most profound impact of space science upon astronomy could be expected in the decades ahead. Just as the new-found ability to study other planetary bodies than the earth immeasurably broadened the perspectives of the earth sciences, so the ability of the astronomer to observe in all the wavelengths that reached the vicinity of the earth could be expected to strengthen the interplay between theory and experiment in the field of astronomy. By the 1970s the process had already begun, but the full power would doubtless have to wait until astronomers had the benefit of a variety of satellites more powerful than the solar and astronomical satellites of the first decade. In addition to large, precise optical telescopes, which one naturally thought of in the 1950s and early 1960s, there would also have to be specially instrumented spacecraft to pursue the new field of "high-energy astronomy" which leapt into prominence with the early discovery of X-ray sources. One would also need both infrared and radio telescopes in orbit. In short, to make the most of the opportunity that had burst upon the astronomical community, there would have to be established in orbit a rather complete facility consisting not just of a single instrument, but of a set of instruments ranging across the whole observable spectrum.

As for the life sciences, space appeared able to contribute in a variety of ways. One could expose biological specimens—including the crews of manned spacecraft—to the environment of space and observe what happened. But the biologists agreed that the most significant contribution of space science to their discipline could well be in exobiology—the study of extraterrestrial life and the chemical evolution of planets.[8] This subject was subsumed under the study of the solar system, since the evolutional histories of the planets, the kinds of conditions they developed on their surfaces and in their atmospheres, would have much to do with whether life formed on the planet, or with how far a lifeless planet moved toward the formation of life.

Thus, as one moved into the 1970s, although space scientists could take much satisfaction in the wide variety of individual disciplines to which they had been able to contribute, it was the new perspective that brought groups of disciplines together in a common endeavor that was most important.

EXPLORATION OF THE SOLAR SYSTEM

By the end of the 1960s study of the solar system extended from earth to the nearest planets, Mars and Venus; and men had landed on the moon. In the next two years the Apollo astronauts made a searching exploration of the moon, after which Skylab crews turned attention toward earth and the sun. During the early 1970s unmanned spacecraft also added Mercury, the asteroids, and Jupiter with some of its satellites to the list of objects in the solar system that scientists' instruments had been able to reach and observe, and had begun the long trek to Saturn and the outer planets.

Geodesy

Around earth, satellite geodesy continued to advance steadily. The first half-dozen years of geodetic research using space techniques were largely years of preparation for the ultimate coup, that of establishing a worldwide geodetic net referred to a common reference ellipsoid. Year by year geodesists moved steadily toward that goal. By 1970 such a reference system was in use,

at least among the principal experts, and positions relative to the common reference could be given to 15 meters or better. As for earth's gravitational field, good estimates had been obtained for coefficients for the various harmonics in the spherical harmonic expansion of the geopotential up to at least the 19th degree. Geodesists were quick to point out that, in a little more than a decade, scientists had increased the quantitative knowledge of global positioning and of the size and shape of earth by 10 times, and knowledge of the gravity field by 100 times—the same order of improvement as had been achieved in the previous 200 years.[9]

By the mid-1970s another order of magnitude had been realized in positioning techniques, and one could begin to zero in on accuracies sufficient to match variations in mean sea-level height (centimeters) and the very slow movements of tectonic plate motions and continental drift (centimeters per year). Using a combination of satellite techniques, observations on quasars and pulsars with a method called very long baseline interferometry, and extremely accurate clocks (1 part in 10^{16}) that might be developed with superconducting cavities, one could aspire to positional accuracies of several centimeters relative to the reference ellipsoid.

But, as anticipated, geodesy did not remain earthbound, although the first extension to another planetary body was more by chance than otherwise. Tracking of Lunar Orbiters, five of which were put into orbit of the moon before 10 August 1966 and 1 August 1967, showed a peculiar perturbation of the orbital motion over a number of the circular maria of the moon. Analysis soon indicated that these disturbances were probably caused by unusual concentrations of mass in the maria basins. These mascons, as they came to be called, became one of the many puzzles in connection with the evolution of the moon that scientists had to try to explain. Some suggested that they were caused by heavy metallic material from slow-moving iron meteorites that had gouged out the basins where the mascons are now found. Or they might be due to lavas of different densities that filled the basins with the material now observed in the maria. They could be plugs of denser material from a lunar mantle that was shoved upward after the basins had been formed by impacting meteorites. Those making the above suggestions considered the mascons as strong evidence that the moon was quite rigid. But there were other opinions, as John O'Keefe points out:

> This [a rigid moon] is more or less how Urey saw it. What most of the rest of us saw was that the moon was in imperfect isostatic equilibrium like the earth. Apart from the mascons, there was isostatic equilibrium. Kaula pointed out that the earth, like the moon, has important deviations from isostasy, like the Hawaiian mass. Urey . . . overestimated the significance of the "mascons" as indicating a difference between the earth and the moon. He maintained that the moon was more rigid than the earth right up to the time when actual measurements showed that in fact it is less rigid.[10]

Whatever the mascons might turn out to be, however, they were an exciting and clearly significant discovery of the first extension of geodesy into the rest of the solar system.

Atmospheric and Ionospheric Studies

Like geodesy, upper atmospheric research and ionospheric physics continued to build upon the groundwork established earlier. Indeed, to the nonexpert the research and results could easily appear to be more of the same. But to the expert progress in the field was nothing short of phenomenal. Questions that had been uppermost in the experimenter's mind a decade before were quickly answered. It was shown that above 200 kilometers the neutral atmosphere became essentially isothermal, varying with the solar cycle from around 600 kelvins at

sunspot minimum to somewhat more than 2000 K at times of very high solar activity. As had been expected the lighter gases, helium and hydrogen were found to predominate above 600 or 700 km, the outermost portions being largely hydrogen. Above 1000 km the positive ions He^+ and H^+ of helium and atomic hydrogen exceeded the concentrations of atomic oxygen O^+, which was the major ionic constituent of the F_2 region from 150 km to above 800 km.[11] At considerably lower altitudes, the D region of the ionosphere was found to be surprisingly complex. In 1965 experimenters found that D-region ionization below 80 km consisted primarily of hydrated protons. Some years later it was shown that negative ions in the D region tended to form complex hydrated clusters.[12] Most important, the various solar radiations responsible for the ionization of the different layers were completely mapped. As Herbert Friedman of the Naval Research Laboratory put it in 1974:

> The solar spectrum is now known with high resolution over the full electromagnetic range. Spectroheliograms in all wavelengths have revealed the spatial structure over the disk [of the sun] and in the higher levels of the corona. We may conclude that the major features of the electromagnetic inputs to and interactions with the ionosphere are well understood.[13]

One of the most significant findings was that the upper atmosphere and ionosphere could not be considered to be in or even near equilibrium previously a common assumption. For example, temperatures of the electrons in the upper ionosphere were found to be appreciably higher than the temperature of the ambient neutral gas, while the ion temperatures were intermediate between the two.[14] A continuing round of dynamic processes characterized these regions, as solar radiations and the day-night cycle set in motion a complex chain of chemical and physical reactions. Also, in the light of space discoveries, the ionosphere was seen to be a part of a much greater magnetosphere within which one could identify a *plasmapause,* inside of which the hot plasma composing the ionosphere rotated with earth and outside of which particle radiations, no longer corotating with earth, exhibited convective motions induced by the solar wind.[15]

But while such detail may be of interest as answering questions that a decade before still puzzled researchers, their true significance lay in the fact that by the 1970s all known major problems of the high atmosphere and ionosphere had a satisfactory explanation based on sound observational data. From then on research in the upper atmosphere and ionosphere could be regarded as largely a mopping-up operation, the investigation of the finer details of what was going on.

Meteorology

As for meteorology, the story was somewhat different. Sounding rockets provided the means for making measurements at all levels within the atmosphere, and satellites furnished worldwide imaging of cloud systems plus observations of atmospheric radiations and temperature profiles. These data amplified by orders of magnitude the amount of information available to the meteorologist, filling in enormous gaps that had existed over the oceans and uninhabited land areas. But most of the impact of these data was on the forecasting of weather and climatic trends, where their contribution was of inestimable value. In a decade and a half of giant strides in practical meteorology brought about in part by space methods, nothing of a revolutionary nature was contributed to the science of the lower atmosphere. In the mid-1970s atmospheric scientists still had to admit that there had been no breakthroughs attributable to space observations, although a wealth of new information was available and was under continuing intensive study. Most researchers were, however, optimistic that in the years ahead space

data would share with ground-based, balloon, and aircraft measurements in leading ultimately to the breakthroughs in that understanding of the atmosphere needed to provide long-term forecasts of both weather and climate and to predict accurately the place and time of occurrence of severe storms.

If there was no general breakthrough, there were several intriguing contributions from space research. For one thing, as with other areas of the earth sciences, the perspective afforded by satellite imaging was a great stimulus to research. The ability to see and assimilate with ease the distribution and kinds of clouds, the location and nature of weather disturbances, the distribution of vorticity, etc., gave the researcher a new handle on his subject. As one result, tropical meteorology, once regarded as a dead-end field, sprang to life; and scientists began to develop new insights into the relations between the tropics and mid-latitudes.

Most of the sun's radiation lies in the visible wavelengths, which, along with some infrared and ultraviolet, control earth's weather. This portion of the sun's spectrum is remarkably constant over time, although the question of just how constant remains open, and is one of the problems that space methods may help to solve. The short wavelengths, on the other hand, are extremely variable, their changes causing enormous variations within the high atmosphere. A natural question, then, was whether these upper atmospheric and ionospheric changes might not be related to meteorological changes. But, although sudden warmings of the stratosphere appeared to be associated with solar ultraviolet radiations, the general view was that this portion of the solar spectrum, containing less than one one-millionth of the energy carried in the visible wavelengths, could hardly have any significant effect. Yet, after more than a decade of space research, intriguing hints of relationships between upper atmospheric and meteorological activity began to appear.[16] For exam-

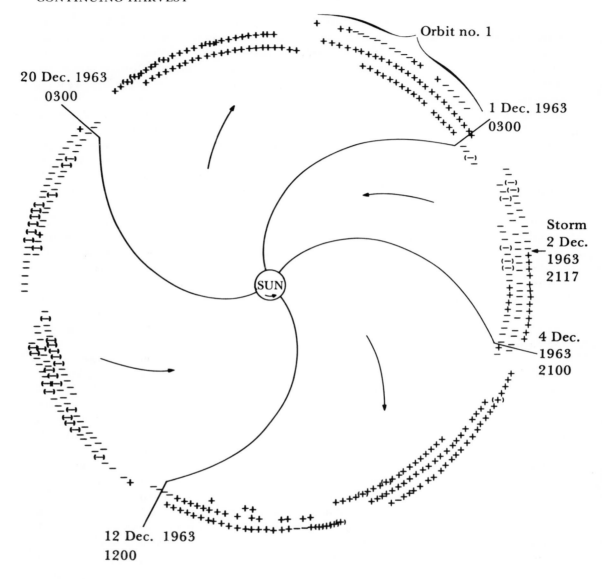

Figure 1. Sector structure of the solar magnetic field. The data are for December 1963 to February 1964. The direction of the average experimental interplanetary magnetic field during three-hour intervals is denoted by plus signs (away from the sun) and minus signs (toward the sun). Parentheses around a plus or minus sign indicate that the field direction fluctuated significantly. The solid lines represent magnetic field lines at sector boundaries. Alois W. Schardt and Albert G. Opp in *Significant Achievements in Space Science, 1965,* NASA SP-136 (1967), p. 42.

ple, particles-and-fields research had shown that the interplanetary medium around the sun was divided into sectors in some of which magnetic fields were directed away from the sun, while in others the magnetic fields were pointed generally toward the sun (Figure 1). The two kinds of sectors alternated with each other in going around the sun.[17] Quite remarkably the boundaries between these sectors appeared to be associated with changes in atmospheric vorticity. Here was a phenomenon that could have a profound significance and the existence of which lent substance to the question of magnetospheric and upper atmospheric relationships to meteorology.

As space scientists were getting a firm grip on the physics and chemistry of earth's upper atmosphere, their attention was simultaneously being drawn toward the planets. What was known about planetary atmospheres had come from the efforts of a small, select group of scientists, mostly astronomers.[18] Even more remote from the astronomers than earth's upper atmosphere had been from the geophysicists, the atmospheres of Mars, Venus, and the other planets taxed the investigators' ingenuity. Gross uncertainties often existed in their estimates of atmospheric properties. As with the earth's upper atmosphere, measurements from space probes promised to eliminate or reduce many of the uncertainties.

The Soviet Venera spacecraft in 1970 and 1972 removed any doubt that the ground-level pressure of Venus's atmosphere was about 100 times that of earth, and the Jet Propulsion Laboratory's Mariners showed that Mars's atmosphere was roughly one percent that of earth. The atmospheres of both Venus and Mars were established as primarily carbon dioxide, as had been concluded from ground-based observations. As radio astronomers had already shown, the surface of Venus was confirmed to be in the vicinity of 700° K and fairly steady, while that of Mars was somewhat colder than earth's and varied appreciably with the seasons. Whereas both earth and Mars rotated rapidly with approximately the same periods, ground-based radar measurements showed Venus to turn very slowly—once every 243 days—in the opposite direction to that of the other planets.[19]

From the point of view of comparative planetology, the relations between Venus, earth, and Mars were ideal. Earth was clearly intermediate between the two others in many respects, and many scientists felt that a detailed study of all three should be of special benefit in understanding earth. An example of the kind of interplay that was possible was furnished by the study of the role of halogens in the atmosphere of Venus. The investigations led to the suspicion that chlorine produced in earth's stratosphere from the exhausts of Space Shuttle launches or from freon used at the ground in aerosol sprays might dangerously deplete the ozone layer, which was known to shield earth's surface from lethal ultraviolet rays of the sun. In a similar manner, when Venus's atmosphere was found to exhibit a single circulation cell, global in extent,[20] it was recognized that a careful study of this special example could yield important insights into the terrestrial atmosphere, where numerous circulation cells interact on a rapidly rotating globe. The clouds on Venus had long been a mystery, in which stratospheric aerosols now appeared to play a key role. The unraveling of the precise role of aerosols in the Venus atmosphere would certainly benefit studies of chemical contamination of earth's atmosphere. At the other end of the scale, the role of dust storms in the thin Martian atmosphere could lend an important additional perspective to the role of dust in modifying earth's climate. On a much grander scale, as Pioneer spacecraft passed by Jupiter in 1973 and 1974 it was learned that the famous red spot[21] was a huge hurricane large enough to engulf three earths. What might be learned from the Jupiter hurricane about atmospheric dynamics that could be applied to the case of earth remained to be seen.

The first planetary ionosphere other than earth's to be detected experimentally was that of Mars. By observing the influence of the planet's atmosphere on radio signals from *Mariner 4* as those signals traversed the planet's atmosphere just before the spacecraft was occulted by the planet, it was possible to obtain an estimate of electron density as a function of altitude. The ionosphere, which was observed at a period of minimum sunspot activity, was somewhat less developed than expected from terrestrial analogies. Later, at times of greater activity, *Mariner 6, 7,* and *9* revealed a slightly more intense ionosphere, showing a noticeable dependence on the solar cycle. The same sort of occultation experiment on *Mariner 5* (1967) gave electron density profiles for Venus, both dayside and nightside. The nightside ionosphere was almost two orders of magnitude less intense than the daytime ionosphere, which showed a distinctly higher electron density than that of Mars.[22] Jupiter was shown to have a well-developed ionosphere.[23]

Thus, during the 1960s, while satisfactory answers were being obtained for all the known, major problems of earth's high atmosphere, a good start was made on the investigation of the atmospheres and ionospheres of other planets.

Magnetospheric Physics

A genuine product of the space age, magnetospheric research also moved on apace. During the first half-dozen years the principal task was for the experimenters to produce an accurate description of the magnetosphere, although once the existence of the radiation belts, a terrestrial magnetosphere, and the solar wind had been revealed by actual observation in space, a host of theorists vied with each other to devise explanations. Indeed, Eugene Parker's seminal paper on the solar wind antedated the detection of the wind and met with a great deal of flak until his critics were silenced by the space observations. The situation was typical for science; often many competing theories produce a continuing argument which no one can win until specific measurements become available to weed out those theories that do not fit the data. After the initial years of discovery and survey, however, the main action shifted to the theorists,[24] although the experimentalists continued to amass additional data from both satellites and space probes.[25]

At the end of the 1960s the theorists could explain many features of sun-earth relations, the interplanetary medium, and the magnetosphere, but a large number of fundamental questions remained to be resolved. To the layman a schematic picture of the magnetosphere drawn at the end of the decade might look much like one produced a half-dozen years earlier. But the expert would read into that diagram a new collection of rather subtle questions that still had to be answered before one could claim to have a thorough understanding of magnetospheric physics.[26] The initial reaction to the discovery that earth's magnetic field generated a huge bow shock in the rapidly moving solar wind was to apply hydrodynamical theory. The general shape and position of the bow shock and the magnetopause could be understood from magnetohydrodynamical principles. Also, it appeared that one might explain the sudden commencement and initial phase of magnetic storms in a straightforward way. But the theory could not explain why the solar wind appeared to apply a surface drag to earth's field, pulling some of the field lines out into the long geomagnetic tail that was a spectacular aspect of the magnetosphere. Instad one would expect the magnetospheric cavity to close off in a teardrop shape.

To resolve some of these difficulties attention turned to the idea that the solar wind was a collisionless plasma, and the bow shock a collisionless phenomenon. Since under this assumption particle-to-particle collisions would be negligible, one had to seek the cause of the bow shock in cooperative field effects, such as interactions between electrostatic fields of the charged particles and magnetic field components perpendicular to the direction in which the gas velocity changed. As the 1970s opened, a great deal of study was going into collisionless shocks, particularly turbulent shocks, which observations showed earth's bow shock to be.

Other problems that required attention were the wide range of geomagnetic activity, the acceleration of particles within the magnetosphere, the production of the auroras, and the formation of the geomagnetic tail. In connection with these matters, the idea that parallel and opposite magnetic fields might merge and annihilate each other aroused stormy debate. One could point to cases in which such a process might be important. For example, if the field in the solar wind had a southward component when it struck earth's magnetic field at the nose of the magnetosphere, where the terrestrial field would be northward, merging and annihilation might take place. Or merging could occur when field lines in the interplanetary medium happened to be essentially parallel to field lines in the magnetospheric tail.

With such problems the magnetospheric physicists had an agenda that would keep them amply occupied during the 1970s. Moreover, early in that decade they got their first look at another planetary magnetosphere—that of Jupiter.[27] As the fascinating complexity of earth's magnetosphere and its important role in sun-earth relationships had unfolded, physicists had immediately thought of the possibility of other planetary magnetospheres. It was known from observation of Jovian radiations that Jupiter had a very strong magnetic field.[28] As a consequence no one doubted that the first spacecraft to reach the giant of the solar system would encounter a well defined magnetosphere.

Until instruments could be sent to the nearer planets, the question remained open as to whether they also had magnetospheres. The first Soviet Luniks showed that the moon had very little magnetic field, an observation that was confirmed repeatedly in later US missions.[29] Although some magnetism was found in the lunar crust, it became quite clear that the moon did not have a poloidal field such as the dipolar field of earth. Accordingly there was no lunar magnetosphere analogous to that of earth. Similarly the Mariner that flew by Venus in 1962 could detect no magnetic field, nor could *Mariner 4* when it reached Mars in 1965.[30] But, surprisingly, Mercury when reached in 1974 turned out to have an appreciable field.[31]

These circumstances provided the space scientists an opportunity for comparative studies in sun-planetary relations, an opportunity that in the early 1970s was still largely unexploited. Earth provided the case of a planet with a sizable atmosphere and a strong magnetosphere, both of which were involved in intricate ways in the processes by which the sun exerted its influence on the planet. At the other extreme the moon provided the case of a body with neither magnetosphere nor atmosphere, so that solar radiations impinged directly upon the lunar surface. For Venus the sun's particle radiations struck the atmosphere directly, unmodified by a magnetosphere; but the extremely dense atmosphere shielded the planet's surface completely from these radiations. Mars, with its thin atmosphere but also without a magnetosphere was intermediate between the moon and Venus. Mercury, on the other hand, provided an example of a planet with a magnetosphere but no atmosphere. How such a magnetosphere would differ from earth's, which was continually interacting with the planet's atmosphere and ionosphere, was an interesting subject for investigation, one that doubtless would be explored over the years ahead.[32]

In December 1973 *Pioneer 10* reached Jupiter, followed a year later by *Pioneer 11*. Their instruments revealed a huge magnetosphere reaching 7.6 million kilometers into space. Within the magnetosphere were radiation belts ten thousand to a million times as intense as those of earth. Jupiter's magnetosphere reached well beyond the orbits of its four largest satellites, those observed first by Galileo centuries ago. The innermost satellite, Io, appeared to interact strongly with the magnetospheric radiations.[33] But to pursue these fascinating investigations any further would go beyond the scope of this article.[34]

Planetology

The final topic of this section concerns the planetary bodies themselves. While investigation of earth's atmosphere, ionosphere, and magnetosphere—and related solar studies—were naturally the first areas of research in the space science program, they had only limited appeal to the layman. As exciting as these challenges were to the researchers, the average person could hardly relate to himself the magnetohydrodynamic concepts, terrestrial ring currents, or complex photochemical reactions in the ionosphere. But the investigation of the moon and planets was different. Here in pictures one could see landscapes and clouds—often strange, to be sure, but landscapes and clouds nevertheless. One could envision spacecraft orbiting a planet or landing on its surface and could identify personally with astronauts stepping onto the bleak and desolate moon. As a consequence NASA had little difficulty in capturing and holding a widespread interest in this aspect of the space science program.

The exploration of the moon and planets began with the Soviet Luna flights in 1959. From that time on, every year at least one mission to the moon or a planet was attempted by the United States or the Soviet Union. The American assault on the moon began with Pioneers, followed by Rangers, then the soft-landing Surveyors. In the summer of 1966 the first of five Lunar Orbiters began the task of mapping almost the entire surface of the moon. Even an Explorer was injected into lunar orbit to study the space environment around the moon. The climax was reached when the Apollo missions began manned exploration of the moon with the orbital flight of *Apollo 8* in December 1968 and the first manned landing in July 1969. While Apollo was in progress the Soviet Union conducted a series of sophisticated unmanned lunar missions that included circumlunar flights of Zond

spacecraft, which were successfully recovered with pictures they had taken of the moon. More advanced Luna spacecraft soft-landed on the moon, carrying roving vehicles to investigate the lunar surface *in situ* and radio the information to Soviet stations on earth, and in some cases to send samples of lunar soil to earth for investigation in the laboratory. The success of the Soviet unmanned rovers and sampling missions sparked an intense debate between the scientists and NASA, many of the scientists feeling that the unmanned approach to the study of the moon was the wiser, and by far the more economical.

Criticism was blunted, however, by the tremendous success of Apollo. The astronauts brought back hundreds of kilograms of lunar rocks and soil from six different locations, the analysis and study of which quickly engaged the attention of hundreds of scientists throughout the United States and around the world. In addition to collecting lunar samples, the astronauts also set up nuclear-powered geophysical laboratories instrumented with seismometers, magnetometers, plasma and pressure gauges, instruments to measure the flow of heat from the moon's interior, and laser corner reflectors for geodetic measurements from earth. The geophysical stations operated for many years after the astronauts had left, radioing back volumes of information on the moon's environment and its seismicity. Twice satellites were left behind in lunar orbit for lunar geodesy and to make extended chemical analyses of the lunar surface material from observations of the short-wavelength radiations of the moon.

The United States claimed the first success in planetary exploration when *Mariner 2,* launched in the summer of 1962, passed by Venus the following December, probing the clouds, estimating planetary temperatures, measuring the charged particle environment of the planet, and looking for a magnetic field.[35] In 1965 *Mariner 4* flew by Mars to take 21 pictures, covering about one percent of the planet's surface. Then *Mariner 5* visited Venus, this time getting substantially more data on the atmosphere, including estimates of the ionosphere. It was back to Mars in 1969 with *Mariner 6* and *7,* which returned some 200 pictures of the surface along with a variety of other measurements. *Mariner 9* went into orbit around the Red Planet in November 1971 at a time when the planet was almost completely obscured by a global dust storm. During the next month and a half the spacecraft monitored the clearing of the dust storm, which itself provided much interesting information about the planet. After that the spacecraft's cameras were devoted to the first complete mapping ever achieved of another planet—*Mariner 9* returned 7,329 pictures of Mars and its two satellites, which permitted drawing up complete topographical maps showing the true nature of the markings that had so long puzzled astronomers.[36]

During the 1960s the Soviet Union had also set its sights on the planets. In fact, its attempts appreciably outnumbered those of the United States, since the USSR seized virtually every favorable opportunity to try a launching. But early Soviet planetary endeavors were about as dismal as America's early lunar tries. Not until a Venera spacecraft in 1970 succeeded in penetrating the atmosphere of Venus, returning data on the composition and structure of the atmosphere, did fortune smile on these planetary attempts. In December 1970 *Venera 7* landed and returned data from the surface of Venus; *Venera 8* followed suit in July 1972, for 50 minutes sending back surface data and analyses of the soil of Venus. Less fortunate, however, were the two Soviet Mars landers which, like *Mariner 9,* also arrived at the planet during the great dust storm of 1971. The storm that provided the Mariner with an unexpected opportunity to observe the dynamics of the planet's atmosphere may have been the cause of the Soviet spacecraft's failure to land successfully.[37]

In November 1973, *Mariner 10* left on a journey that would take it first by Venus and then on to Mercury, where the spacecraft arrived in March 1974 taking pictures and making a variety of other measurements. Having completed its first Mercury mission, *Mariner 10* was redirected by briefly firing its rockets so that the spacecraft would visit Mercury again in September of 1974. By visiting Mercury several times, *Mariner 10* provided the scientists with the equivalent of several planetary missions for little more than the price of one.[38] Also, with the visit to Mercury, scientists at long last had close looks at all of the inner planets of the solar system, including the two satellites of Mars.

The result of all these space probe missions was the accumulation of volumes of data on the moon and near planets, illuminated with thousands of highly detailed pictures. The photo resolutions exceeded by orders of magnitude what had been possible through telescopes. When Rangers crashed into the moon the closeup pictures sent back just before the impact were a thousandfold more detailed than the best telescopic pictures previously available (Figure 2). After landing on the surface with its television cameras, Surveyor afforded another thousandfold increase in resolution, revealing the granular structure of the lunar soil and a considerable amount of information on the texture of lunar rocks (Figure 3). In the laboratory Apollo samples put the moon's surface under the microscope, as it were (Figure 4). As for the planets no detail at all had been available before on the surface characteristics of distant Mercury or clouded Venus. Some *Mariner 10* pictures afforded better resolution for Mercury than earth-based telescopes had previously given for the moon (Figures 5-6). Ultraviolet photos of Venus from passing spacecraft showed a great deal of structure in the atmospheric circulation that was hitherto unobservable (Figure 7), while radar measurements from earth penetrated the clouds to reveal a rough, cratered topography.[39] For Mars, the indistinct markings observable

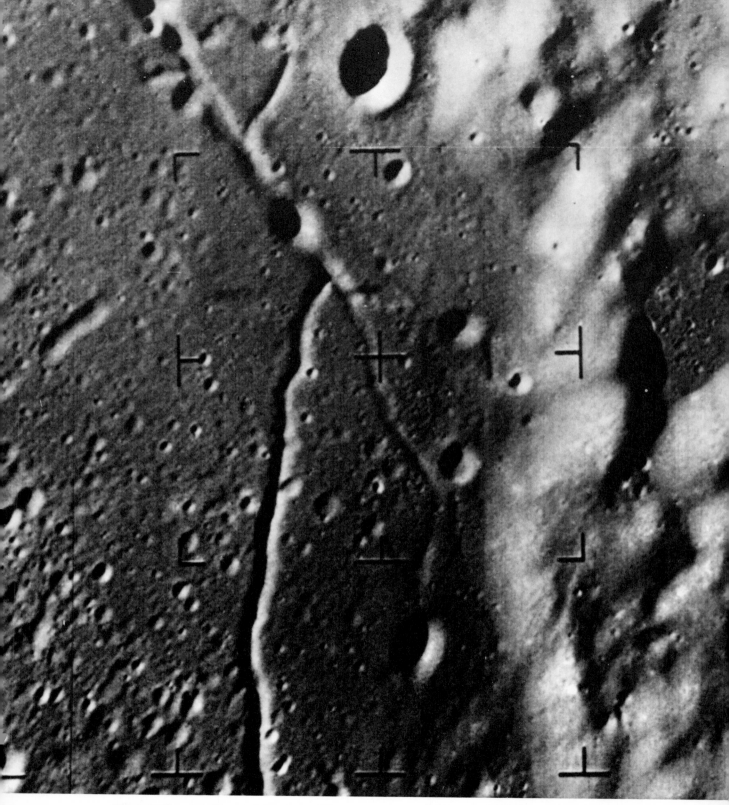

Figure 2. Ranger photos of the moon. The closeup picture, above, of the crater Alphonsus floor taken by *Ranger 9* in 1965, shows detail not available in telescopic pictures taken from earth. *Ranger VIII and IX*, JPL Tech. Rpt. 32-800, pt. 2 (15 March 1966), p. 353, fig. 7.

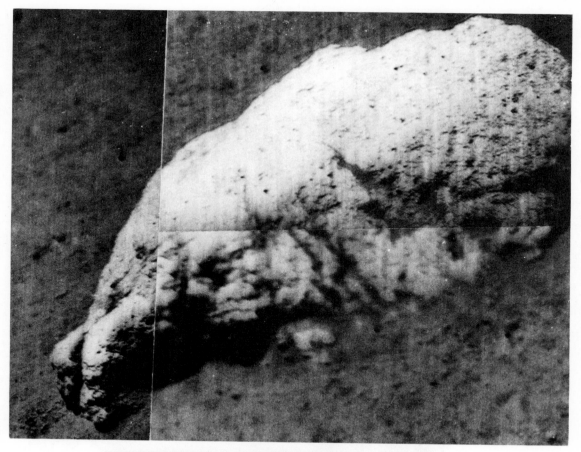

Figures 3-4. Surveyor photos of the moon, top right and right. Sitting on the moon's surface, *Surveyor 1* in 1966 provided a millionfold increase in resolution over that of earth-based pictures. *Surveyor Program Results,* NASA SP-184 (1969), p. 259, fig. 7-42.

from earth were replaced with sufficient detail to show craters, volcanoes, rifts, flow channels, apparently alluvial deposits, sand dunes, and structure in the ice caps (Figures 8-15). Added to such pictures, data on planetary radiations in the infrared and ultraviolet; surface temperatures; atmospheric temperatures, pressures, and composition (when there was an atmosphere); and charge densities in the ionosphere (when there was an ionosphere)—this wealth of information completely revitalized the field of planetary studies, which had long been quiescent for lack of new data. By the early 1970s comparative planetology was well under way, although one must hasten to add that the task ahead of understanding the origin and evolution of the planets was one of decades, not merely months or years.

Nevertheless, progress was rapid. Much was learned about the mineralogy and petrology of the moon, and by extrapolation probably about the other terrestrial planets. Radioactive dating of lunar specimens led to the conclusion that the moon is probably some 4.6 billion years old, an age consistent with the ages of meteorites and the presumed age of earth. The moon was found to be highly differentiated; that is, the lunar materials, through total or partial melting, had separated into different collections of minerals and rock types. The maria were mainly basalt, similar to but significantly different from the rocks of the ocean basins on earth. In contrast the lunar highlands were rich in anorthosite, a rock consisting mainly of the feldspar calcium aluminum silicate. Both maria and highlands were much cratered—as could already be seen from earth—and one could now see that the crater sizes extended down to the very small, showing that the moon had been bombarded by very small particles as well as by very large

Figure 5. Thin section of *Apollo 16* lunar sample 67075. Samples of lunar material brought to earth made it possible to examine minerals of the moon under a microscope. The area of the section photographed is 2.3 mm long.

objects. The entire surface was covered with fine fragments and soil, broken rocks and rubble—crustal material chopped up by the cratering process. A considerable amount of glass was found, some of it in coatings splashed onto other rocks, much of it in the form of tiny glass beads of a variety of colors dispersed through the soil.

Reading through the record imprinted on the lunar surface one could bit by bit piece together the course of the moon's evolution. Contrary to expectations of many, before the unmanned and manned space missions showed differently, the present lunar surface was not a virginal record of conditions that existed at the time of the moon's formation. Instead one could discern a continual, sometimes violent process of evolution. Either at the time of formation or very shortly thereafter the moon's crust was molten to a considerable depth. Presumably during this phase the differentiation of lunar materials took place, producing the light-colored, feldspar-rich highlands. After solidifying about 4.4 billion years ago, the highlands were bombarded for hundreds of millions of years, probably by material left over from the formation of the moon and planets. This process produced the cratered topography of the highlands still visible today. Between 4.1 and 3.9 billion years ago the bombardment of the moon became cataclysmically violent, with asteroid-sized meteorites gouging out the great basins, hundreds and even thousands of kilometers across (Figure 16). Then, as radioactive dating of lunar materials showed, between 3.8 and 3.1 billion years ago a series of eruptions of basaltic lavas filled the basins to form the maria, or dark regions of the moon clearly visible from earth. By 3 billion years ago the violent evolution of the moon had come to an end. Cratering and "gardening" of the lunar surface continued, but at about the rates observed today, most of the impacting particles being micrometeor and grain sized, occasionally cobble sized, with the very large impacts exceedingly rare.[40]

166

Figures 6-7. Photomosaics of Mercury. Eighteen pictures, taken at 42-second intervals, were enhanced by computer at the Jet Propulsion Laboratory and combined into this mosaic. The pictures were taken from *Mariner 10* during 13 minutes when the spacecraft was 200,000 kilometers and six hours away from Mercury on its approach to the planet, 29 March 1974. About two-thirds of the portion of Mercury seen in this mosaic is in the southern hemisphere. Largest of the craters are about 200 kilometers in diameter. Illumination is from the left. The semicircle of cratered mountains in the left half of the mosaic forms the boundary for the largest basin on Mercury seen by *Mariner 10*. The basin is near a subsolar point when the planet is at perihelion, leading investigators to suggest the name Caloris for it. The ring of mountains is 1300 kilometers in diameter and up to 2 kilometers high. The basin floor consists of severely fractured and ridged plains.

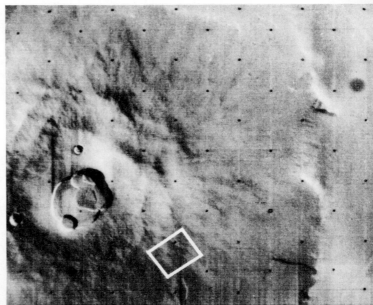

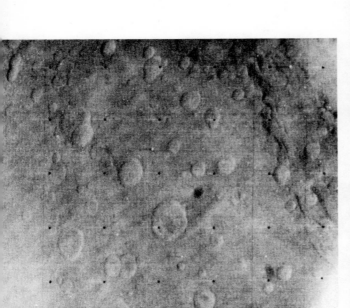

Figure 9. Craters on Mars. The slightly darker region to the right is Hellespontus; the lighter region to the left, Hellas. Apparently an escarpment forms the boundary between the two regions. The area of the photo, taken by *Mariner 7* in 1969, is 720 by 960 kilometers.

Figure 8. Cloud structure on Venus. The structure reveals the pattern of atmospheric circulation on the planet. The picture was taken in ultraviolet light by *Mariner 10* cameras on 6 February 1974.

Figure 10. Volcano on Mars. Olympus Mons was photographed 7 January 1972 by *Mariner 9*. The top photo shows an area 435 by 655 kilometers. The lower, high-resolution photo, corresponding to the inscribed rectangle at the top, shows details of lava flow down the mountain side.

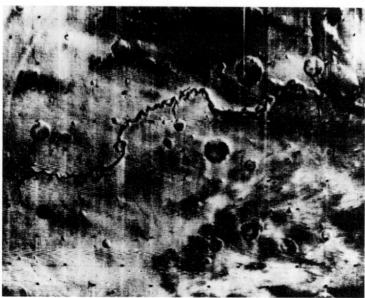

Figure 11. Rifting on Mars. The superimposed outline of the United States emphasizes how the great canyon on Mars dwarfs the Grand Canyon on earth. The photomosaic was made from *Mariner 9* pictures obtained in several weeks of photographic mapping of the planet. The area covered reaches from $-30°$ to $+30°$ latitude and from $18°$ to $140°$ longitude.

Figure 12. Flow channel on Mars. This valley, some 400 kilometers long, resembles a giant version of an arroyo on earth.

Figure 13. Evidence of water flow on Mars, at right. Braided channels at $-6°$ latitude, associated with Vallis Mangala in the Amazonis region of Mars near longitude $150°$. Such features are common to sediments deposited during meandering stream flow. Each of the two frames in this composite covers about 30 by 40 kilometers.

170

171

Figure 14. Sand dunes on Mars, above. The presence of sand and dust on Mars was dramatically emphasized by the great dust cloud that enveloped the planet as *Mariner 9* approached. After the cloud of dust had settled, deposits of sand were observed to shift about from photos of the same region taken at different times.

Figure 15. Ice cap on Mars, below. The photo shows the northern hemisphere of Mars from the polar cap to a few degrees south of the equator. At the stage shown, the ice cap is shrinking during late Martian spring.

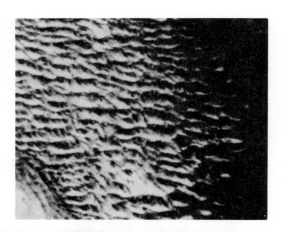

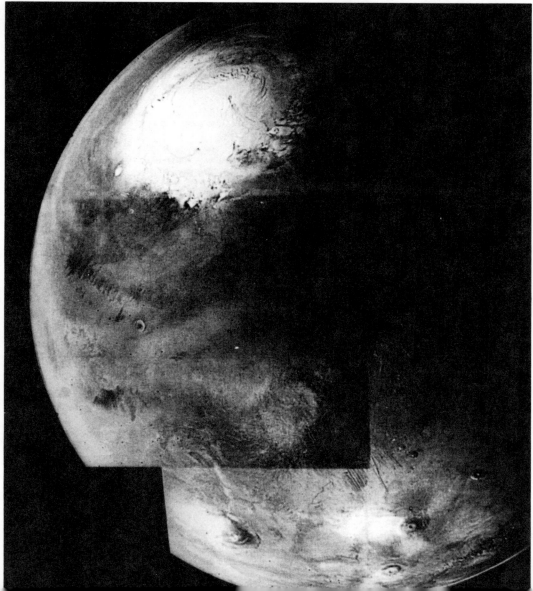

Figure 16. The moon. The great maria, like Mare
Imbrium in the upper hemisphere, were probably gouged out by huge meteorites, and subsequently filled by extensive flows of dark
basaltic lava.

The moon observed by Apollo instruments was very quiet. The lack of any substantial organized magnetic field suggested that there was no molten core, but the presence of magnetism in lunar rocks indicated that there might have been a core at some time in the past—on the assumption that a liquid iron core was required to generate the lunar field that magnetized the lunar rocks. Seismometer data showed that the moon was at least partly molten below 800 km and suggested a small, possibly iron, core a few hundred km in radius. The energy released by moonquakes detected by the Apollo seismometers was nine orders of magnitude—a billion times—less than that released by earthquakes over a similar period of time, which, however, seemed consistent with the relative sizes of earth and moon. The seismic data yielded the picture of a lunar crust 50 to 60 km thick, four times the average thickness of earth's crust, underlain by a mantle solid at the top but partially melted toward the bottom (Figure 17). The very thick crust could account for the lack of any recent volcanic activity.[41]

Before these investigations it was common to think of the moon as a small body and to suppose that small bodies would remain cold and essentially unchanged throughout their histories. Now it was clear that bodies the size of the moon, and even smaller ones, whether formed in the molten state or melted after formation, undergo a substantial evolution. This conclusion was borne out further by data from the other planets. Mercury appeared even more cratered than the moon. There was widespread evidence of lava flows on the planet. Large cracks and long scarps were visible in the *Mariner 10* pictures. There was no doubt that Mercury underwent a great deal of evolution after its formation.[42]

The evidence of activity on Mars was even more striking. After a period of discouragement for scientists when the *Mariner 4* pictures appeared to show a dead, moonlike planet, the pictures of *Mariner 6* and 7 revived interest, and those of *Mariner 9* aroused excitement. Those pictures showed huge volcanoes, one of them—Olympus Mons—twice the diameter of the largest known volcanic structure on earth, namely, the big island of Hawaii. A great deal of the Martian surface was cratered, indicating an ordeal of bombardment like that experienced by the moon and Mercury. Some of the surface was smooth and featureless, indicating a process of filling in as with blowing, drifting sands. A huge rift, comparable in length to the width of the United States, indicated considerable tectonic activity. Numerous channels hundreds of kilometers long looked as though they might have been produced by flowing water. Consistent with this observation were the frequent formations that looked like alluvial fans produced by the deltas of terrestrial rivers or sedimentary deposits of meandering streams. While the variable frost cover observed in the north and south polar caps was shown to be solid carbon dioxide, a substantial base of frozen water was also found. Various estimates suggested that a considerable amount of water must have outgassed from the planet over time, and if past conditions on the planet were just right there could have been ponds and rivers. But the question of the role of water in the evolution of the planet remained unsolved.[43]

It was very clear that Mars is an active planet, by no means dead, as some had prematurely concluded. Some investigators thought they could detect in the marked differences between the cratered highlands of the planet and the volcanic provinces the suggestion of an incipient separation into individual tectonic plates as on earth, a picture not generally accepted by students of Mars. Paul Lowman, however, was led to conclude that the evidence was piling up that earthlike planetary bodies would follow similar courses of evolution.[44] In the larger bodies like earth, the rates of evolution would be faster and the duration longer than for the smaller planets. Of the inner

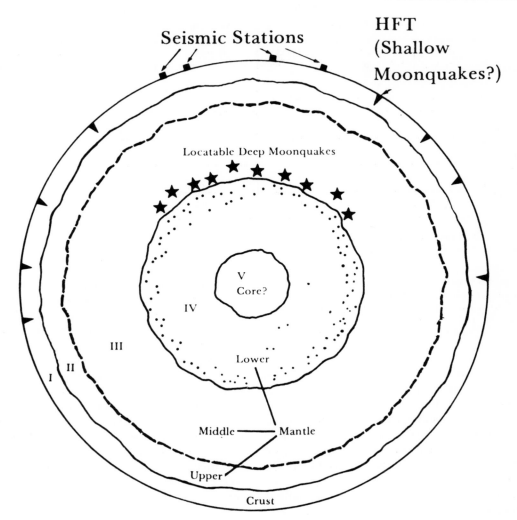

Seismic Stations

HFT
(Shallow
Moonquakes?)

Locatable Deep Moonquakes

V
Core?

IV

III

II

I

Lower

Middle —— Mantle

Upper

Crust

HFT = High-frequency teleseismic event.

Figure 17. Structure of the moon. As on earth, seismic data reveal a great deal about internal structure. Y. Nakamura, G. Latham, et al., in *Proceedings of the Seventh Lunar Science Conference,* ed. R. B. Merrill (New York: Pergamon Press, 1976), pp. 3113-21; reproduced by permission of Gary V. Latham.

planets, earth, still vigorously active, was most advanced in its evolutionary course (Figure 18). Venus might be in a comparable stage, but no life evolved there to convert the carbon dioxide atmosphere to one with large amounts of oxygen. Mars was following the course taken by earth, but was well behind, only now approaching the tectonic plate stage. The moon and Mercury had long since run the course of their evolution, which terminated well before a tectonic plate stage.

This picture, although consistent with much of the data, could hardly be regarded as more than tentative. It would have to pass the test of further observations and measurement, and stiff debate. But one satisfying

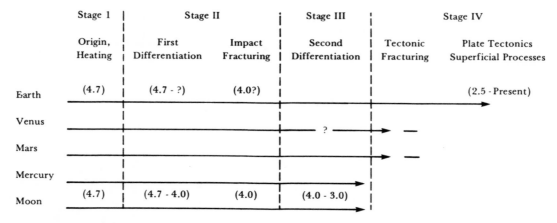

	Stage 1	Stage II		Stage III	Stage IV	
	Origin, Heating	First Differentiation	Impact Fracturing	Second Differentiation	Tectonic Fracturing	Plate Tectonics Superficial Processes
Earth	(4.7)	(4.7 - ?)	(4.0?)			(2.5 - Present)
Venus				?		—
Mars						—
Mercury						
Moon	(4.7)	(4.7 - 4.0)	(4.0)	(4.0 - 3.0)		

Ages ( ) in billions of years before present;
refer to peak of events; events may overlap.

Figure 18. Crustal evolution in silicate planets. Evidence suggests that earthlike planets all follow similar courses of evolution. Paul D. Lowman, Jr., in *Journal of Geology* 84 (Jan. 1976): 2, fig. 1; reproduced courtesy of Dr. Lowman.

feature was the emphasis the theory gave to the kinship of the planets with each other. As theorists had pointed out, if the planets did form from the material of a solar nebula left over after the creation of the sun, then their individual characteristics should depend to a considerable extent on their distances from the sun (Figure 19). Near the sun, where the nebular material would be heated to rather high temperatures by the sun's radiations, one could expect to find planets composed primarily of materials that condense at high temperatures, the silicates and other rock-forming minerals. Moreover, the densities of the planets could be estimated according to distance from the sun by considering what compounds were likely to form at the temperature to be expected at Mercury's distance, which at the distance of Venus, which in the vicinity of earth, etc. From what was known of the inner planets, they did indeed fit such a picture.

As for the outer planets, one would expect them to consist of large quantities of the lighter substances—hydrogen, helium, ammonia, methane—which could condense out of the solar cloud only at the low temperatures that would exist so far from the sun. Qualitatively, the outer planets also fitted this picture, but quantitatively there were discrepancies. To develop the true state of affairs in proper perspective, an intensive investigation of the outer planets was called for, and was on the agenda for the 1970s and 1980s. The investigation got off to an exciting start with the visit of *Pioneer 10* to Jupiter in 1973. It was clear that an exciting period in planetary exploration lay ahead as scientists began to amass data on the atmospheres, ionospheres, and magnetospheres of these strange worlds. While these planets themselves would be quite different from the terrestrial planets, their satellites could be expected to resemble the latter in many ways. Moreover, as many persons pointed out, the satellite systems of Jupiter and Saturn might turn out to be very much like miniature solar systems, particularly the satellites that formed along with the parent planet rather than being captured later. Supporting this view was the early discovery from the Pioneer observations that the four regular satellites of Jupiter decreased in density with increasing distance from the planet, as though they had formed from a cloud of gas and dust that was hotter near

176

the planet than it was farther away.[45] The opportunities for important research seemed endless.

Finally there was the question of extraterrestrial life. Space research on fundamental biology was early divided into two areas: (1) the study of terrestrial life forms under the conditions of space and spaceflight, and (2) exobiology. The latter came to mean the search for extraterrestrial life and its study in comparison with earth life. Most scientists considered the chance of finding life elsewhere in the solar system to be minute,

Figure 19. Origin of solar system planets. The higher temperature materials, like silicates, condense nearer the sun; the more volatile substances, farther away.

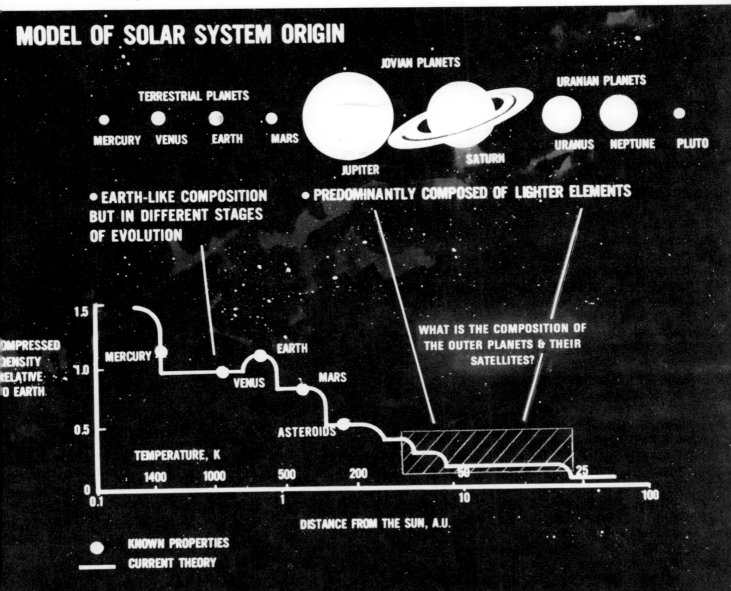

but it was universally agreed that the discovery of such life would be a tremendously important event. Thus, while recognizing the unlikelihood of finding extraterrestrial life, many considered that the poential implications offset the small chance of finding any, and accordingly devoted considerable time to studying in the laboratory the chemical and biological processes that seemed most likely to have been part of the formation of life. They sought out earth forms that could live under extremely harsh conditions—like arid deserts, the brines of the Great Salt Lake, or the bitter cold Antarctic—and paid special attention to them. And they devised experiments to probe the Martian soil for the kinds of life forms deemed most likely to be there.

But, after a decade and a half, the problem of life on other planets remained open. No life was found on the moon, nor was there any evidence that life had ever existed there. A careful search was made for carbon, since earth life is carbon-based. In lunar samples a few hundred parts per million were found, but most of this carbon was brought in by the solar wind. Of the few tens of parts per million that were native to the moon, none appeared to derive from life processes.[46]

Nor was any life found on Mars, even though the two Viking spacecraft with their samplers and automated laboratories were set down in 1976 in areas where once there might have been quite a bit of water.[47] Still, the subject could hardly be called closed. If life had been found, that would have settled the question. But that life was not found in two tiny spots on Mars did not prove that there was no life on the planet. So, although the first attempts were disappointing, it could be assumed that future missions to Mars would pursue the question further. Nor would it be likely that exobiological research would be confined to the Red Planet. At the very least, one would expect that, as scientists studied the chemical evolution of the planets and their satellites, they would

keep the question of the formation and evolution of life in mind.

While the foregoing has touched upon but a few of the results accruing from the exploration of the solar system, still the reader should be able to derive some insight into the impact that space science was having upon the earth and planetary sciences. For one thing the study of the solar system was revitalized after a long period of relative inactivity. Second, lunar and planetary science became an important aspect of geoscience, attracting large numbers of researchers. Third, the new perspective afforded by space observations gave an immeasurable boost to comparative planetology, a field that made great strides during its first 15 years. Nevertheless, no one doubted that in the mid-1970s comparative planetology still looked forward to its most productive years.

All of which leads to the usual question. If space science was having such a profound impact on earth and planetary sciences, was space science producing a scientific revolution in the field? In the broad sense, no. But, viewed strictly from within the discipline there were indeed numerous revolutionary changes. Much new information was accumulated, permitting the theorist to deal in a realistic way with topics about which one could only speculate before. Many had to relinquish pet ideas about the nature of the lunar surface or the markings on Mars. Proponents of a cold moon were faced with incontrovertible evidence of extensive lunar melting. No pristine lunar surface was to be found; instead a substantial evolution had marked the moon's first one and a half billion years. Far from being an inert planet, Mars turned out to be highly active. Of course, in different aspects of the subject many investigators had been on the right track. Noted astronomers R.B. Baldwin and E.J. Öpik has correctly anticipated that many of the features of Mars were due to craters.[48] Gerard Kuiper had been sure that volcanism was important on the moon, as he explained many times to the author. Thomas Gold had been certain that the lunar surface would contain a great deal of fine dust. Yet, no one had succeeded in putting the separate pieces together in satisfactory fashion. Thus, the most revolutionary aspect of space science contributions to the earth and planetary sciences was probably in helping to develop an integrated picture of the moon and near planets. This was an enormous expansion of horizons, an expansion that could be expected to continue with each new planetary mission.

Investigation of the Universe

As space scientists were busily altering the complexion of solar system research, space methods were also profoundly affecting the investigation of the universe. Here space science could contribute in a number of ways to solar physics, galactic and metagalactic astronomy, and cosmology, including a search for gravitational waves, observations to determine whether the strength of gravity was changing with time, and studies of the nature of relativity. But the contribution of space techniques to these areas was qualitatively different from those in the planetary sciences. Whereas rockets and spacecraft could carry instruments and sometimes the observers themselves to the moon and planets to observe the phenomena of interest close at hand, this was not possible in astronomy. The stars and galaxies would remain as remote as before, and even the sun would continue to be a distant object extremely difficult to approach even with automated spacecraft because of the tremendous heat and destructive radiation.

The connection between the scientists and the objects of study would continue to be the various radiations coming from the observed to the observer. But rockets and satellites would increase the variety of radiations that the scientist could study by lifting telescopes and other instruments above earth's atmosphere, which was transparent only in the visible and some of the radio wavelengths. This extension of the observable

spectrum proved to be as fruitful to the prober of the universe as were the lunar and planetary probes to the student of the solar system.

Rocket astronomy began in 1946 when sounding rockets were outfitted with spectrographs to record the spectrum of the sun in hitherto hidden ultraviolet wavelengths. In 1948 X-ray fluxes were detected in the upper atmosphere, after which rocket investigations of the sun ranged over both ultraviolet and X-ray wavelengths. Inevitably experimenters turned their instruments on the skies, and when they did various ultraviolet sources were found. In 1956 the Naval Research Laboratory group found some celestial fluxes that might have been X-rays, but the real significance of X-rays for astronomy had to await more sensitive instruments that did not become available until the early 1960s.

In the meantime sounding rocket research on the sun's radiations moved on apace. Investigators from a number of institutions continued to amass detail on the sun's spectrum in the near and far ultraviolet, which was important in understanding the quiet sun and normal sun-earth relationships. But the real excitement proved to be with the X-rays. It was these, rather than the ultraviolet wavelengths, that came into prominence with high solar activity. When satellites came into being, they were put to use in making long-term, detailed measurements of the sun's spectrum in all wavelengths. On 7 March 1962 the first of NASA's Orbiting Solar Observatories went into orbit, to be followed by a series with steadily improving instrumentation. The Naval Research Laboratory built and launched a series of Solrad satellites, intentionally less complex than the OSOs, to provide a continuous monitoring of the sun in key wavelengths. But, while satellites came into prominence in the 1960s, sounding rockets, some of them launched at times of solar eclipse, continued to yield important results. In fact, some scientists felt that the most significant work on the sun came from sounding rockets rather than from the far more expensive satellites.

NASA's Orbiting Solar Observatories continued into the 1970s, the first one of the decade being *OSO 7,* launched on 29 September 1971. An important event for solar research was the launching of Skylab in 1973. In this space laboratory astronauts studied the sun intensively using a special telescope mount built for the purpose. Although the high cost of Skylab's solar mission in dollars and time to prepare and conduct the experiments was distressing to many of the scientists, nevertheless the results were extremely important for solar physics, some of them providing solutions to long unsolved problems.

Sounding rocket experiments were also fruitful in stellar astronomy. Perhaps the most significant event in rocket and satellite astronomy occurred when American Science and Engineering experimenters, with an Aerobee rocket flown on 12 June 1962, discovered the first X-ray sources outside the solar system to be clearly identified as such. As will be seen later, this discovery proved to be of profound significance to modern astronomy. During the 1960s sounding rockets continued to search for and gather information on these strange sources, but progress was slow. Long-term observations with more precise instruments were needed, a need NASA was much too slow in supplying. The breakthrough came with the launching of NASA's first Small Astronomy Satellite, *Explorer 42,* on 12 December 1970.

During the 1960s the course of ultraviolet astronomy from satellites also proceeded slowly. The principal satellite designed for such studies, NASA's Orbiting Astronomical Observatory, proved difficult to bring into being, and it was not until the end of the decade that *OAO 2* (7 December 1968), the first successful astronomical observatory, went into orbit. It took another four years to get *OAO 3* aloft (21 August 1972). From

OAO observations the Smithsonian Astrophysical Observatory compiled the first complete ultraviolet map of the sky, issuing the results in the form of a catalog for use by astronomers.[49] With *OAO 3*, which had been named *Copernicus* in honor of that dauntless pioneer in scientific thought, Princeton University experimenters obtained a number of significant results. They showed that, while hydrogen in the interstellar medium was almost entirely in atomic form, most interstellar clouds had an abundance of neutral hydrogen molecules, the relative abundance being consistent with a balance between the catalytic formation of H_2 on grains of material and the competing dissociation of the gas by absorption of light. Much of the galactic disk was found to be occupied by a hot coronal gas at half a million kelvins, with a hydrogen density of one particle per liter. The Princeton workers also observed that the relative abundance of the different chemical elements in the interstellar gas was what would be expected if, starting with a mixture of elements in the ratios found in the overall cosmic abundance, the materials of high-condensation temperatures had already condensed out to form small solid particles or dust grains. Finally, flowing from most very hot stars were stellar winds of some thousands of kilometers per second.

Other experiments were also on the OAOs, some of them concerning X-rays and gamma rays, and the Orbiting Astronomical Observatories were clearly proving very fruitful. Yet one could detect the feeling that OAO was a bit out of step. The satellite had been sufficiently difficult to construct that it had delayed satellite optical (visible plus ultraviolet) astronomy for about a decade, whereas a series of cheaper, simpler satellites could have kept research moving while work on a larger instrument proceeded. Also, now that it had come, OAO was well behind both existing telescope technology and current needs. For most of the problems of greatest concern to the optical astronomers, a larger aperture (2.5 to 3 m), a more precise telescope was required. As agitation developed for the construction of a Space Shuttle it was quickly realized that one of the things that the Shuttle could do ideally was to launch such an instrument into orbit and service it throughout the years. The launching of such a telescope became one of the prime scientific missions for the Shuttle.

But in the 1970s the circumstances surrounding astronomy had changed. Whereas in 1959 and 1960 the most important tasks for satellites had appeared to be in ultraviolet astronomy, in the late 1960s and early 1970s both ground-based and space research had changed the picture. Now the high-energy end of the spectrum—X-rays and gamma rays—was the center of attention for many astronomers, particularly for the large number of physicists who had moved into the field of astronomy. As a result a number of NASA's sounding rockets and small astronomy satellites were devoted to this area of research. In addition the agency began to plan for outfitting and launching a series of multiton satellites for X-ray and gamma-ray astronomy—to be called High Energy Astronomical Observatories. The importance of this work in the eyes of scientists was shown by the fact that Britain, the Netherlands, and the European Space Agency all instrumented satellites of their own for high-energy astronomy work.

The result of all the research with sounding rockets and satellites was an outpouring of data, not obtainable from the ground, at a time when ground-based astronomy was each year turning up new, exciting, often unexplainable discoveries. The quasars had extremely large red shifts in their spectra, suggesting that they were among the most remote of objects observed, yet if they were as remote as indicated, then they would have to be emitting energy at rates that defied explanation. Strange galaxies appeared to have violent nuclei, emitting unexplainable quantities of energy. The discovery of pulsars introduced the neutron star to the scene. Radio galaxies gave evidence of cata-

clysmic explosions in their centers. The rocket and satellite could not have appeared at a more propitious time.

As with the exploration of the solar system, the flood of new data and information was far beyond what could be covered in a brief summary like the present one. To keep within bounds, it is necessary to illustrate the impact of space science upon the field by means of selected examples. Two will be given: X-ray astronomy and some of the contributions of space science to solar physics.

X-ray Astronomy

Once stars are born the major part of their evolution can be followed in the optical wavelengths—that is, the visible and ultraviolet. For this reason most attention was directed at launching space astronomy in the direction of ultraviolet studies. Very little thought was given to the higher energy wavelengths, although these were proving to be extremely informative about the sun, particularly about solar activity. But there were a few who thought that one ought to look for celestial X-ray sources. Perhaps the most insistent was Bruno Rossi, professor of physics at the Massachusetts Institute of Technology. As Rossi later said, while he had not been in a position to predict specific phenomena like the X-ray sources that were eventually discovered, he had had "a subconscious trust . . . in the inexhaustible wealth of nature, a wealth that goes far beyond the imagination of man."[50] Moreover, there was a very compelling reason to try to look at the universe in the very short wavelengths. Much has been made of the fact that sounding rockets and satellites gave experimenters their first opportunity to look at all the wavelengths that reached the top of the atmosphere. But, like the atmosphere, interstellar space also had its windows and opacities, and not all wavelengths emitted in the depths of space could reach the vicinity of earth. While the interstellar medium was quite transparent to wavelengths all the way from radio waves through much of the near ultraviolet, at and below 1216 A absorption by hydrogen, the most abundant gas in space, cut off radiations from distant objects. Farther into the ultraviolet, absorption decreased again, rising once more when the absorption lines of helium, also abundant in space, were encountered in the far ultraviolet. Not until the X-ray region was observed could one again see (with instruments, not with the eyes) deep into the galaxy. The existence of that window in the spectrum was an important reason for sounding out the possibilities of X-ray astronomy. One could study the universe only in the wavelengths that were available for observation, and all known windows ought to be exploited.

Rossi urged his ideas upon Martin Annis and Riccardo Giacconi of American Science and Engineering, who were enthusiastically receptive.[51] Quick calculations showed that X-ray intensities one might expect from galactic sources would be well below the detection limit for existing instruments— doubtless explaining why Naval Research Laboratory searches for X-ray sources had not found any. In 1960 NASA provided support to Giacconi and his colleagues and they prepared to fly sufficiently sensitive instruments in sounding rockets.

After one failure, the group succeeded in getting an Aerobee rocket to an altitude of 230 kilometers, on 12 June 1962. Although the planned objective of the flight was to look for solar-induced X-radiation from the moon, that objective was completely eclipsed by the excitement of detecting an object in the sky that was apparently emitting X-rays at a rate many, many orders of magnitude greater than the sun. The sheer intensity of the source gave one pause, and the experimenters spent a considerable time reviewing their results before announcing them in late summer. In two additional flights, October 1962 and June 1963, Giacconi's group was able to confirm the original findings and discover additional sources, again detecting a strong isolated source near the center of the galaxy.[52] There was also a diffuse isotropic background which one supposed could be extragalactic in origin.

Within a few months of Giacconi's original announcement, the Naval Research Laboratory experimenters had confirmed the existence of these sources by independent observations. In April 1963 the NRL group made an important contribution by pinpointing the source near the galactic center more precisely. Since it lay in the constellation Scorpius, the source was named Sco X-1. NRL also detected a somewhat weaker source, about 1/10 the strength of Sco X-1, in the general vicinity of the Crab Nebula.[53] During the rest of the decade additional sources were discovered until by 1970 several dozen were known. Efforts were made to discover just what these strange objects were. In particular, it was felt especially important to identify the sources with objects already known in the visible or radio wavelengths. The reason for wanting to tie X-ray sources in with known objects was quite simple. It seemed clear that the extremely intense X-radiation from these sources had to be connected with the basic energetics of the objects. It could be very important then to compare the X-radiation with that in other wavelengths, a comparison that might better reveal the true nature of the phenomenon.

As an illustration consider the Crab Nebula.[54] It is believed to be the remnant of a supernova explosion. The material of the nebula consists of the debris ejected by the exploding star into the surrounding space. At present this debris, which is expanding at about 1000 kilometers per second, fills a roughly ellipsoidal volume with the major diameter about six light years—where a light year is 9500 trillion km. Radiation from the Crab Nebula was observed to be strongly polarized across the whole spectrum, probably resulting from electrons revolving about the lines of force of a magnetic field. At the densities that appeared to exist in the cloud, the lower-energy electrons that would produce the radio wavelengths observed could last for thousands of years, but those responsible for optical wavelengths would be depleted in a few hundred years and those producing the X-rays in less than a year. Hence there had to be a continuous resupplying of energy to maintain the observed radiations.

This remained a mystery until the discovery of a pulsar in the Crab. The end product of a supernova explosion was expected to be a very dense neutron star, in which a mass comparable to that of the sun would be compressed into a ball about 10 km in diameter. There were reasons to believe that the pulsations from which a pulsar took its name were generated by the rapid spinning of the star. When it was noted that the period of the Crab pulsar was lengthening at the rate of about one part in 2000 per year, a possible source for the resupply of energy to the nebula leapt to mind. The slowdown of the neutron star's rotation corresponded to a considerable amount of energy, and calculations soon showed that the amount was adequate to provide the energy being released by the nebular gases. One speculation had it that the spinning magnetic field of the pulsar accelerated gases to relativistic speeds at which they could escape from the star's magnetosphere into the nebula, carrying their newly acquired energy with them.

This example illustrates the importance of being able to observe and visualize the object that was radiating the X-rays. As a consequence there was a continuing effort to identify the other X-ray sources with known objects. But, except for Sco X-1, the efforts were unsuccessful. As for Sco X-1, in an experiment in March 1966, Giacconi's group showed that the angular size of the source could not exceed 20 arcseconds. The new positional data were communicated to workers at the Tokyo Observatory and at Mt. Wilson and Palomar Observatories. On 17-18 June 1966 the Tokyo observers found a blue star of 13th magnitude, which further study identified with the X-ray source. This result confirmed that stars existed that emit-

ted 1000 times as much power in X-rays as in the visible wavelengths. As Giacconi wrote: "Sco X-1 was a type of stellar object radically different from any previously known and whose existence had not and could not be foreseen on the basis of observations in the visible and radio."[55]

But that was as far as one appeared able to go with the sounding rockets. Longer-term observations with very sensitive detectors were required to advance the field further. These were supplied by *Explorer 42*. Launched from an Italian platform in the Indian Ocean off the coast of Kenya on 12 December 1970, Kenyan Independence Day, the satellite was at once named *Uhuru*, the Swahili word for independence. *Uhuru* provided the breakthrough astronomers were looking for. When issued sometime afterward, the *Uhuru* catalog listed 161 X-ray sources. It was clear that the strongest sources had to have X-ray luminosities at least 1000 times the luminosity of the sun. Even the weakest X-ray source was 20 times more intense than the meter-band radiation from the strongest radio source. Thirty-four sources had been identified with known objects. Of the sources known to lie within the galaxy, one was at the galactic center, seven were supernova remnants, and six were binary stars. Outside the galaxy were sources associated with ordinary galaxies, giant radio galaxies, Seyfert galaxies, galactic clusters, and quasars. And much of the diffuse background X-radiation had been shown to come from outside the local galaxy.[56]

One of the most exciting results from X-ray astronomy came from the realization that some at least of the sources were binary stars—two stars revolving around each other (Figure 20)—in which one of the companions was a very dense star, either: (1) a white dwarf, a mass comparable to that of the sun compressed into a sphere about the size of earth; or (2) a neutron star, of mass exceding 1.4 solar masses, compressed into a ball about 10 km in radius; or (3) a black hole in space. A white dwarf is the end product of a

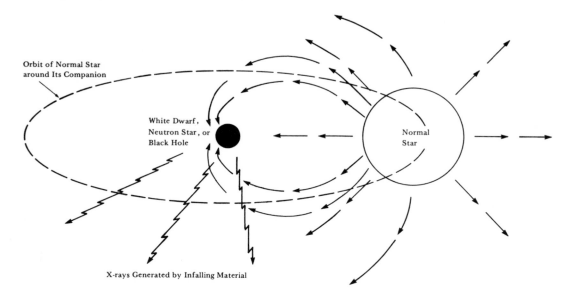

Orbit of Normal Star
around Its Companion

White Dwarf,
Neutron Star, or
Black Hole

Normal
Star

X-rays Generated by Infalling Material

star about the size of the sun, after it has used up its nuclear fuel and can no longer avoid collapsing under gravity to a planetary size. The neutron star is the end product of a massive star that, after burning up its nuclear fuel, undergoes a violent implosion caused by gravity, following which a great deal of the star's material rebounds from the implosion to be blown out in a supernova explosion, leaving behind an extremely dense object consisting of neutrons. But if the residual mass after the supernova explosion is greater than a certain critical value, the gravitational contraction of the star does not stop even at the neutron star stage. Instead the star continues to contract indefinitely, pulling the matter tighter and tighter together until the object disappears into a deep gravitational well out of which neither matter nor electromagnetic radiation can escape because of the intense gravitational fields there. Hence the name "black hole."

The binary nature of some of the X-ray objects could be deduced from the doppler shifts in the light from the ordinary companion, the shift being toward the blue as the star moved toward the observer in its orbit, and toward the red as the star moved away.

Figure 20. Binary x-ray star. Material from the larger, much less dense companion is drawn toward the extremely dense smaller companion and accelerated to velocities sufficient to produce X-rays by collision with the ambient gases.

If the stars eclipsed each other the binary nature would show up in a periodical disappearance of the X-rays when the emitter was hidden by the other star and reappearance when the emitter emerged from eclipse.

After careful study astronomers finally concluded that the X-rays were generated by material from the ordinary companion's being pulled into the gravitational well of the degenerate star. If the gravitational attraction were sufficiently strong, then the gas would be accelerated to such velocities that the gas would emit X-rays as particles collided. It seemed that white dwarfs would not provide sufficient gravity to accomplish this, so one was left with the conclusion that the degenerate companion in binary X-ray sources was either a neutron star or a black hole in space. In most cases it appeared that the companion was a neutron star, but the source Cyg X-1, in the constellation Cygnus, could be a black hole. If so, it was the first such object to be detected in the universe.[57]

The possibility that a black hole had at last been discovered emphasized the fundamental importance to astronomy of the new field of X-ray and gamma-ray astronomy. Gradually scientists had begun to talk about their work as *high-energy astronomy,* not only because they were working at the high-energy end of the wavelength spectrum, but more significantly because their observations were showing that throughout the universe extremely violent events were rather common, involving enormous quantities of energy and tremendous rates of energy production. And among these energetic events were those occurring during the last stages of a star's evolution, stages in which neutron stars and black holes were created, with intense X-ray emissions. Speculating on the philosophical implications, Giacconi showed some of the excitement that scientists felt over what the future might bring:

> The existence of a black hole in the X-ray binary Cyg X-1, has profound implications for all of astronomy. Once one such object is shown to exist, then this immediately raises the possibility that many more may be present in all kinds of different astrophysical settings. Supermassive black holes may exist at the center of active galaxies . . . and explain the very large energy emission from objects such as quasars. Small black holes of masses [very much smaller than the mass of the sun] may have been created at the instant of the primeval explosion. . . . In black holes matter has returned to condition similar to the primordial state from which the Universe was created. The potential scientific and intellectual returns from this research are clearly staggering.[58]

Should one then conclude that rocket and satellite astronomy had by the early 1970s generated a scientific revolution in the field of astronomy? The answer may well be yes, although many of the strange concepts that were being dealt with had been considered decades before.[59] In any event, it is probably too early to make the case. Certainly these topics concerning the interplay of energy and matter on a cosmological scale, are fundamental; and if anywhere in the space program one might expect a scientific revolution to emerge, it would be here. But it should also be noted that if any such revolution is to arise, it would almost certainly come from a cooperation between ground-based and space astronomy.

Solar Physics

The sun was a most important target of space science investigations for at least two reasons. First, the sun's radiation supports life on earth and controls the behavior of the atmosphere. For meteorology it was important to know the sun's spectrum in the visible, infrared, and near-ultraviolet wavelengths. To understand the various physical processes occurring in the upper atmosphere, a detailed knowledge of the solar spectrum in the ultraviolet and X-ray wavelengths was essential. The reader will recall the overriding importance that S.K. Mitra, in his 1947 assessment of major upperatmospheric problems, gave to learning about the electromagnetic radiations from the sun. For this reason many sounding rocket experimenters devoted much of their time to photographing and analyzing the solar spectrum both within and beyond the atmosphere. Finally, with the discovery of the magnetosphere and the solar wind the importance of the particle radiations from the sun for sun-earth relationships became apparent. Thus, solar physics was of central importance in the exploration of the solar system.

But the sun was important also to astronomy, to the investigation of the universe. Although an average star, unspectacular in comparison with many of the strange objects that astronomers were uncovering in their probing of the cosmos, nevertheless it is a star, and it is close by. The next star, Proxima Centauri, is 4.3 light-years (400 trillion kilometers) away, while most of the stars in the galaxy are many tens of thousands of light-years distant. Stars in other galaxies are millions and even billions of light-years from earth. So the sun afforded the only opportunity for scientists to study stellar physics with a model that could be observed in great detail.

Because of its nearness and its importance, astronomers amassed a great deal of data and theory about the sun in the years before rockets and satellites.[60] What they learned came almost entirely from observations in the visible, with only occasional glimpses from mountain tops and balloons at the shorter wavelengths. But, as space observations showed, much of solar activity, particularly that associated with the sunspot cycle, solar flares, and the corona involved the short wavelengths in essential ways. Sounding rocket and satellite measurements were, accordingly, able to round out the picture in important ways.

To understand the significance of these contributions, a brief summary of the principal features of the sun may be helpful.[61] The visible disk of the sun is called the photosphere (Figure 21). It is a very thin layer of one to several hundred kilometers thickness, from which comes most of the radiation one sees on earth. The effective temperature of the solar disk is about 5800° K. Above the photosphere lies what may be called the solar atmosphere; below it, the solar interior.

The sun's energy is generated in the interior, from the nuclear burning of hydrogen to form helium in a central core of about one-fourth the solar radius and one-half the solar mass. Here, at temperatures of 15×10^{6}° K, some 99 percent of the sun's radiated energy is released. This energy diffuses outward from the core, colliding repeatedly with the hydrogen and helium of the sun, being absorbed and reemitted many times before reaching the surface. In this process the individual photon energies continually decrease, changing the radiation from gamma rays to X-rays to ultraviolet light and finally to visible light as it emerges from the sun's surface.

From the core of the sun to near the surface, energy is thus transported mainly by radiation. But toward the surface, between roughly 0.8 and 0.9 of the solar radius, convection becomes the principal mode of transporting energy toward the surface. The existence of the convection zone is evidenced by the mottled appearance of the photosphere in high-resolution photographs. This mottling, or granulation as it is called, consists of cells of about 1800-km diameter which last about ten minutes on the average and which are thought to be associated with turbulent convection just beneath the photosphere. A larger scale system of surface motion—20 times the size of the granulation cells, called supergranulation—is believed to be much more deeply rooted in the convection zone.

Analogous to earth's upper atmosphere is the chromosphere, or upper atmosphere of the sun. The chromosphere overlies the photosphere and is about 2500 km thick. While the density drop across the photosphere is less than an order of magnitude, density in the chromosphere decreases by four orders of magnitude from the top of the photosphere to the top of the chromosphere.

Not long ago, photospheric and chromospheric temperatures were extremely puzzling to astronomers, dropping from about 6600° K at the base of the photosphere to around 4300° at the base of the chromosphere, and then rising through the chromosphere, at first slowly but then very steeply to between 500,000 and 1,000,000° K at the top. This temperature curve posed a problem, for it was assumed that the corona derived its heat from the chromosphere, yet that would imply that heat was flowing from

Figure 21. Structure of the sun (not to scale). There is a complex interplay among the different regions of the sun. Edward G. Gibson, *The Quiet Sun*, NASA SP-303 (1973), p. 11, fig. 2-3.

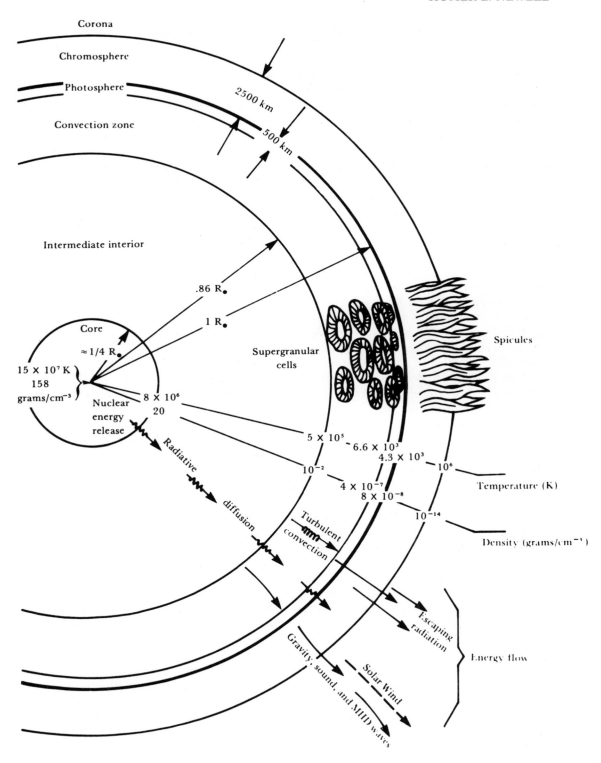

a colder region to a hotter one, contrary to the laws of thermodynamics. As late as 1972 Leo Goldberg, director of Kitt Peak National Observatory, pointed to this phenomenon as "the most important unsolved mystery surrounding the quiet sun."[62]

Above the chromosphere lies the corona, the sun's exosphere. Here 1,000,000° K temperatures prevail, and an important problem facing the solar physicist was to explain how the corona gets its energy. Although the corona is extremely hot and very active, its density is so very low that it is not normally visible from the ground, where it is completely obscured by scattered sunlight in the earth's atmosphere. Only during solar eclipses, with the moon blocking out the sun's disk, could the astronomer get a good look at the entire corona. One of the benefits of rockets and satellites was to permit carrying coronagraphs above the light-scattering atmosphere where the corona could be seen even in the absence of a solar eclipse.

Much of solar physics concerns the interplay among the different regions of the sun. This interplay, however, can be followed only in terms of its effect upon the radiations emitted from those regions. For this reason, one of the first tasks of the astronomer was to obtain good spectra of the sun and their variation with time. Regions from which radiations of highly ionized atoms came would be hot regions, and temperatures could be estimated. The magnetic field intensities, for example in sunspots, could be estimated from the splitting of lines emitted within the field. If a cooler gas overlay a hotter, similar gas, the cooler gas would absorb some of the light emitted by the hotter one. This would produce reversals in the emission lines of the hotter gas, generating the famous Fraunhofer lines of the solar spectrum discovered in the 19th century. By piecing together information of this kind, the locations of different gases relative to each other and their temperatures could be determined. Changes in magnetic fields that occurred in association with solar activity,

such as the appearance of solar flares, could be followed. Changes were important, since there were strong indications that magnetic fields were the source of much of the energy in solar flares.

These techniques were, of course, applicable in the visible wavelengths and were employed to the fullest by the ground-based astronomer. The space astronomer simply provided an additional handle on things by furnishing spectral data in the ultraviolet and X-ray wavelengths. And these data began to accumulate from the very earliest sounding rocket flights. Year by year, flight by flight, they were added to until by the end of the decade the solar spectrum was known in great detail from visible through the ultraviolet wavelengths and into the X-rays.[63]

A powerful technique for study of the sun is that of imaging the sun in a single line; for example, the red line emitted by hydrogen known as hydrogen alpha. In such spectroheliograms, as they are called, one can see the structure and activity of the sun associated with that line. Spectroheliograms taken in hydrogen, calcium, and other lines in the visible have long been an effective tool for the study of solar activity. Members of the Naval Research Laboratory group pioneered the use of this technique in space astronomy, where it was possible to get spectroheliograms in both the ultraviolet and X-rays.[64] These, taken with photographs in the visible, gave a powerful means of discerning and analyzing active regions on the sun. Sequences of such images taken over many days, or at intervals of 27 days, the solar rotation period, permitted one to follow the evolution of flares and other features on the sun. It was in this sort of imaging that Skylab was particularly fruitful.

During the decade and a half that was climaxed by the Skylab solar observations, solar physics progressed rapidly, advanced by a combination of ground-based and space astronomy. In the shorter wavelengths the sun was found to be extremely patchy, a patchiness that extended into the visible wavelengths as well.[65] The sequence of events in a solar flare could be followed in wavelengths all the way from X-rays through the ultraviolet and visible into the radio-wave region, and related to motions of electrons and protons associated with the flare.[66] Contrary to previous expectations fostered by ground-based pictures during solar eclipses, the corona turned out to be not even nearly homogeneous. X-ray images of the corona especially showed a great deal of structure. Quite surprising were large-scale dark regions of the corona—which came to be called dark holes—and hundreds of coronal bright spots. The holes appeared to be devoid of hot matter and to be associated with diverging magnetic field lines of a single polarity. If the magnetic field lines were open, these holes could be a source of particles in the solar wind.[67]

The bright spots were observed to be uniformly distributed over the solar disk. They were typically about 20,000 km in diameter, and of a temperature about 1,600,000° K. They appeared to be magnetically confined, and one speculated that they might be an important link in the explanation of the sun's magnetic field.[68]

The interlocking features of the lower solar atmosphere and the corona visible in satellite images of the sun provided hints as to how the sun might heat the corona to the extreme temperatures that were observed. Gravity waves and acoustical waves might carry energy upward from the convective regions below the photosphere into the corona. This explanation would remove the mystery of the steep temperature curve in the chromosphere. It would not be the chromosphere that was heating the corona in violation of the laws of thermodynamics. On the contrary, the corona, heated by energy from within the sun, would itself be heating the top of the chromosphere.

The importance of rocket and satellite solar astronomy lay in the integrated attack that the researcher could now make in seeking to understand the nearest star, an integrated attack made possible by opening up the window that the earth's atmosphere had so long kept shut. It was an importance attested to by the large numbers of solar physicists who bent to the task of assimilating the new wealth of data.

By the end of the 1960s the early years of space science were well behind. More than a dozen disciplines and subdisciplines had found sounding rockets and spacecraft to be powerful tools for scientific research. Thousands of investigators turned to these tools to help solve important problems. Moreover, while the disciplines to which the new tools could contribute were many and varied, there was a clearly discernible melding of groups of disciplines into two major fields: the exploration of the solar system and the investigation of the universe. The pursuit of these two main objectives would grow in intensity as space science moved into the 1970s—in spite of fears prevalent in the late 1960s that support for space science was waning. The new decade would witness the scientific missions of Apollo to the moon, the remarkable solar astronomy from Skylab, breakthroughs in X-ray astronomy, and the serious start of a survey of all the important bodies of the solar system. It was eminently clear that space scientists would be important clients of the Space Shuttle, which was intended to introduce a new era in space activities. Because of their accomplishments, the scientists could legitimately ask that the Shuttle be tailored as much to their requirements as to other space needs.

NOTES

1. By way of illustration, see Ivan I. Mueller, *Introduction to Satellite Geodesy* (New York: Frederick Ungar Publishing Co., 1964); William M. Kaula, *Theory of Satellite Geodesy* (Waltham, Mass.: Blaisdell Publishing Co., 1966); G. Mamikunian and M.H. Briggs, eds., *Current Aspects of Exobiology* (Pasadena: Jet Propulsion Laboratory, 1965); Colin S. Pittendrigh et al., eds., *Biology and the Exploration of Mars* (Washington: National Academy of Sciences, 1966); Elie A. Shneour and Eric A. Otteson, compilers, *Extraterrestrial Life: An Anthology and Bibliography* (Washington: National Academy of Sciences, 1966); Robert J. Mackin, Jr., and Marcia Neugebauer, eds., *The Solar Wind* (Pasadena: Jet Propulsion Laboratory, 1966); Wilmot N. Hess and Gilbert D. Mead, eds., *Introduction to Space Science* (New York: Gordon and Breach, 2d ed., 1968); Donald J. Williams and Gilbert D. Mead, eds., *Magnetospheric Physics* (Washington: American Geophysical Union, 1969); Siegfried J. Bauer, *Physics of Planetary Ionospheres* (New York, Heidelberg, Berlin: Springer-Verlag, 1973); Edward G. Gibson, *The Quiet Sun*, NASA SP-303 (Washington, 1973); Riccardo Giacconi and Herbert Gursky, eds., *X-ray Astronomy* (Dordrecht-Holland: D. Reidel Publishing Co., 1974); Stuart Ross Taylor, *Lunar Science: A Post-Apollo View* (New York: Pergamon Press, 1975); Nicholas M. Short, *Planetary Geology* (Englewood Cliffs, N.J.: Prentice-Hall, 1975); R. Grant Athay, *The Solar Chromosphere and Corona: Quiet Sun* (Dordrecht-Holland: D. Reidel Publishing Co., 1976).

2. Hess and Mead, eds. *Introduction to Space Science:* Lloyd V. Berkner and Hugh Odishaw, eds., *Science in Space* (New York: McGraw-Hill Book Co., 1961).

3. Homer E. Newell, "A New Laboratory—How to Work in It," address before American Physical Society, Washington, 29 Apr. 1965.

4. Newell, "NASA's Space Science and Applications Program," statement to Senate Committee on Aeronautical and Space Sciences, 20 Apr. 1967.

5. Homer E. Newell and Leonard Jaffe, "Impact of Space Research on Science and Technology," *Science, 157,* 29-39, 1967.

6. George Gamov, *One Two Three—Infinity* (New York: The New American Library, 1947), pp. 253-314; idem, *The Birth and Death of the Sun* (ibid., 1952); Fred Hoyle, *Frontiers of Astronomy* (ibid., 1955); D.W. Sciama, *The Unity of the Universe* (Garden City,.N.Y.: Doubleday & Co., 1959).

7. Giacconi and Gursky, *X-ray Astronomy.*

8. Pittendrigh et al., *Biology and the Exploration of Mars.*

9. Bernard H. Chovitz, "Geodesy," in *Collier's Encyclopedia, 10* (New York: Crowell-Collier Educational Corp., 1972); 629-38.

10. John O'Keefe to Newell, 22 June 1978, comments on draft Newell MS., NF-40; Chovitz, "Geodesy," p. 638; Short, *Planetary Geodesy,* pp. 72-75.

11. Charles Y. Johnson, "Basic E and F Region Aeronomy," presented at Defense Nuclear Agency Symposium, Stanford Research Institute, Aug. 1971; A.P. Willmore, "Exploration of the Ionosphere from Satellites" *Journal of Atmospheric and Terrestrial Physics, 36,* 2255-86, 1974.

12. S.A. Bowhill, "Investigations of the Ionosphere by Space Techniques," *Journal of Atmospheric and Terrestrial Physics, 36,* 2240, 1974.

13. H. Friedman, "Solar Ionizing Radiation," ibid., p. 2252.

14. Willmore, "Exploration of the Ionosphere from Satellites," p. 2279.

15. Bauer, *Physics of Planetary Ionospheres,* p. 6.

16. W.R. Bandeen and S.P. Maran, eds., *Possible Relationships between Solar Activity and Meteorological Phenomena,* GSFC symposium, 7-8 Nov. 1973, NASA SP-366 (Washington, 1975).

17. *Significant Achievements in Space Science 1965,* NASA SP-136, (Washington, 1967), pp. 40-42.

18. Gerard P. Kuiper, ed., *The Atmosperes of the Earth and Planets,* 2d ed. (Chicago: Univ. of Chicago Press, 1952); Gerard P. Kuiper and Barbara Middlehurst eds., *Planets and Satellites* (ibid., 1961).

19. Short, *Planetary Geology,* pp. 249, 282-87.

20. Ibid., p. 285.

21. Ibid., p. 292.

22. Bauer, *Physics of Planetary Ionospheres,* pp. 194-203.

23. Short, *Planetary Geology,* p. 291.

24. Robert L. Carovillano et al., eds., *Physics of the Magnetosphere* (Dordrecht-Holland: D. Reidel Publishing Co., 1968); B.M. McCormac, ed., *Earth's Magnetospheric Processes* (ibid., 1972); idem, *Magnetospheric Physics* (ibid., 1974); V. Formisano, ed., *The Magnetospheres of the Earth and Jupiter* (ibid., 1975); Syuṇ-Ichi Akasofu, *Physics of Magnetospheric Substorms* (ibid., 1977).

25. Formisano, *Magnetospheres of Earth and Jupiter;* K. Knott and B. Battrick, eds., *The Scientific Satellite Programme during the International Magnetospheric Study* (Dordrecht-Holland: D. Reidel Publishing Co., 1976).

26. E.N. Parker, "Solar Wind Interaction with the Geomagnetic Field," in *Magnetospheric Physics,* ed. Williams and Mead, pp. 3-9; idem. "Dynamical Properties of the Magnetosphere," in *Physics of the Magnetosphere,* ed. Carovillano et al., pp. 3-64.

27. Formisano, *Magnetospheres of Earth and Jupiter.*

28. Zdenek Kopal, *The Solar System* (London: Oxford Univ. Press, 1972), p. 15.

29. *Significant Achievements in Planetology 1958-1964*, NASA SP-99 (Washington, 1966), p. 43.

30. Ibid., pp. 44-46; *Significant Achievements in Space Science 1965,* pp. 139-40.

31. N.F. Ness et al., "The Magnetic Field of Mercury," pt. 1, *Journal of Geophysical Research, 80,* 2708-16, 1975.

32. Y.C. Whang, "Magnetospheric Magnetic Field of Mercury," *Journal of Geophysical Research, 82,* 1024-30, 1977.

33. Short, *Planetary Geology,* p. 291.

34. Formisano, *Magnetospheres of Earth and Jupiter.*

35. Jet Propulsion Laboratory Staff, *Mariner Mission to Venus* (New York: McGraw-Hill Book Co., 1963).

36. Short, *Planetary Geology,* pp. 242-81; Stewart A. Collins, *The Mariner 6 and 7 Pictures of Mars,* NASA SP-263 (Washington, 1971): William K. Hartmann and Odell Raper, *The New Mars: The Discoveries of Mariner 9,* NASA SP-337 (Washington, 1974).

37. Short, *Planetary Geology,* p. 284; Hartmann and Raper, *The New Mars.* pp. 38-42.

38. Short, *Planetary Geology,* pp. 284-90.

39. Ibid., p. 285; NASA release 76-153, 10 Sept. 1976.

40. Paul D. Lowman, Jr., "The Geologic Evolution of the Moon," *The Journal of Geology, 80,* 125-26, 1972; Taylor, *Lunar Science:* Short, *Planetary Geology,* pp. 196-240.

41. Yosio Nakamura et al., "Deep Lunar Interior Inferred from Recent Seismic Data," *Geophysical Research Letters* (July 1974), 137-40.

42. Short, *Planetary Geology,* pp. 287-89.

43. Ibid., pp. 255-59.

44. Paul D. Lowman, Jr., "Crustal Evolution in Silicate Planets: Implications for the Origin of Continents," *Journal of Geology, 84,* 1-26, 1976.

45. Short, *Planetary Geology,* p. 292.

46. Ibid., pp. 231-32.

47. *Viking 1: Early Results,* NASA SP-408 (Washington, 1976), pp. 59-63; American Geophysical Union, *Scientific Results of the Viking Project,* collection of reprints from the *Journal of Geophysical Research* (Washington, 1977).

48. Hartmann and Raper, *The New Mars,* p. 65.

49. Robert J. Davis, William A. Deutschman, and Katherine L. Haramundanis, *The Celescope Catalog of Ultraviolet Stellar Observations* (Washington: Smithsonian Institution, 1973).

50. Bruno Rossi, "X-ray Astronomy," in "Discoveries and Interpretations: Studies in Contemporary Scholarship," 2, *Daedalus,* Fall 1977 (issued as vol. 106, no. 4, of the *Proceedings of the Academy of Arts and Sciences*), pp. 37-58.

51. Riccardo Giacconi, introduction to *X-ray Astronomy,* ed. Giacconi and Gursky, p. 6.

52. Ibid., pp. 8-11.

53. Ibid., pp. 11-12.

54. T.A. Chubb and H. Friedman, "Glimpsing the Hidden X-ray Universe," *Astronautics and Aeronautics,* 7 (Mar. 1969), 50-55.

55. Giacconi, introduction to *X-ray Astronomy,* pp. 15-19.

56. Giacconi, "X-ray Sky," chap. in *X-ray Astronomy,* pp. 155-68.

57. Riccardo Giacconi, "Progress in X-ray Astronomy," talk presented at Thirty-fourth Richtmeyer Memorial Lecture of American Association of Physics Teachers, Anaheim, Calif., 30 Jan. 1975. Preprint Series No. 304, Center for Astrophysics, Cambridge, Mass., pp. 8-19.

58. Ibid., p. 21.

59. Gursky and Ruffini, *Neutron Stars, Black Holes and Binary X-ray Sources,* app. 1, pp. 259-317.

60. Henry Norris Russell, Raymond Smith Dugan, and John Quincy Stewart, *Astronomy,* 2 vols. (New York: Ginn and Co., 1927); Edward G. Gibson, *The Quiet Sun,* NASA SP-303 (Washington, 1973).

61. Giuseppe Vaiana and Wallace Tucker, "Solar X-ray Emission," chapter in *X-ray Astronomy,* ed. Giacconi and Gursky, pp. 171-78; Gibson, *The Quiet Sun,* pp. 7-30.

62. Leo Goldberg, in a foreword to Gibson's *The Quiet Sun.*

63. Richard Tousey, "Some Results of Twenty Years of Extreme Ultraviolet Solar Research." *The Astrophysical Journal, 149,* 239-52, 1967, plus plates; Kenneth G. Widing, "Solar Research from Rockets and Satellites," *Astronautics and Aeronautics,* 7 (Mar. 1969), 36-43.

64. Tousey, "Twenty Years of Ultraviolet Solar Research," plates 19-28.

65. Vaiana and Tucker, "Solar X-ray Emission."

66. Ibid., p. 197.

67. Ibid., pp. 183-84.

68. Ibid., pp. 187-90.

NAME INDEX

(Numerals in italic indicate chapter written by individual.)

This book was produced by the Smithsonian Institution Press, Washington, D.C.
Printed by the Book Crafters, Inc., Chelsea, Michigan.
Set in VIP Baskerville by Hodges Typographers, Silver Spring, Maryland
The text paper is seventy-pound Lakewood with Holliston Roxite cover and Multicolor endpapers.
Designed by Dorothy Fall.